GOVERNING CLIMATE MOBILITY IN AFRICA

Explorations of Adaptation in
Ethiopia and Ghana

Edited by
Ninna Nyberg Sørensen,
Lily Salloum Lindegaard and Neil Webster

First published in Great Britain in 2025 by

Bristol University Press
University of Bristol
1–9 Old Park Hill
Bristol
BS2 8BB
UK
t: +44 (0)117 374 6645
e: bup-info@bristol.ac.uk

Details of international sales and distribution partners are available at bristoluniversitypress.co.uk

Editorial selection and matter © the editors 2025; individual chapters © their respective authors 2025

The digital PDF and ePub versions of this title are available open access and distributed under the terms of the Creative Commons Attribution-NonCommercial-NoDerivatives 4.0 International licence (https://creativecommons.org/licenses/by-nc-nd/4.0/) which permits reproduction and distribution for non-commercial use without further permission provided the original work is attributed.

British Library Cataloguing in Publication Data
A catalogue record for this book is available from the British Library

ISBN 978-1-5292-4539-4 paperback
ISBN 978-1-5292-4540-0 ePub
ISBN 978-1-5292-4541-7 ePdf

The right of Ninna Nyberg Sørensen, Lily Salloum Lindegaard and Neil Webster to be identified as editors of this work has been asserted by them in accordance with the Copyright, Designs and Patents Act 1988.

All rights reserved: no part of this publication may be reproduced, stored in a retrieval system, or transmitted in any form or by any means, electronic, mechanical, photocopying, recording, or otherwise without the prior permission of Bristol University Press.

Every reasonable effort has been made to obtain permission to reproduce copyrighted material. If, however, anyone knows of an oversight, please contact the publisher.

The statements and opinions contained within this publication are solely those of the editors and contributors and not of the University of Bristol or Bristol University Press. The University of Bristol and Bristol University Press disclaim responsibility for any injury to persons or property resulting from any material published in this publication.

Bristol University Press works to counter discrimination on grounds of gender,
race, disability, age and sexuality.

Cover design: Andrew Corbett
Front cover image: iStock/wilpunt

Contents

List of Figures and Tables		v
List of Abbreviations		ix
Notes on Contributors		xi
Acknowledgements		xv

1	Introduction: An Outline of the Analytical Approach to 'Governing Climate Mobility' *Ninna Nyberg Sørensen, Lily Salloum Lindegaard, Neil Webster, Joseph Kofi Teye and Zerihun Mohammed*	1

PART I Climate and Land Cover Changes

2	Vegetation Cover Changes in the Eastern and Upper West Regions of Ghana *Joseph Kofi Teye, Philip Prince Kwasi Mantey, Marie Ladekjær Gravesen, Richard Seyram, Lily Salloum Lindegaard, Francis Xavier Jarawura, Nauja Kleist and Shelta Gatsey*	23
3	Land Cover and Land Use Change and Climate Variability: Evidence from Longitudinal Geographic Information System Data in Shashemene and Tehuledere *Kefyalew Sahle Kibret, Zerihun Mohammed, Lily Salloum Lindegaard, Ninna Nyberg Sørensen, Neil Webster and Dessalegn Rahmato*	54

PART II Local Impacts and Adaptation Strategies

4	Assessing Adaptive Capacity to Climate Change and Variability in the Savannah and Forest Agro-Ecological Zones of Ghana *Francis Xavier Jarawura, Joseph Kofi Teye, Lily Salloum Lindegaard and Nauja Kleist*	89
5	Climate Change and Its Implications for Smallholders' Crop Production: Change from Maize to Teff to Haricot Beans in the Shashemene District, Ethiopia *Busha Teshome and Zerihun Mohammed*	114

6	Khat Cultivation and Climate Change in Tehuledere, South Wollo *Dessalegn Rahmato*	132
7	'They Are Supposed to Stay Home': Examining the Politics of Adaptation and Climate (Im)mobility in Northern Ghana *Lily Salloum Lindegaard, Francis Xavier Jarawura and Nauja Kleist*	150
8	Short-Distance Migration as an Adaptation Strategy to the Impacts of Climate Change: The Case of the Shashemene District *Zerihun Mohammed*	170

PART III Gendered Mobility Practices

9	Gender Dimensions of Climate Change-Related Migration in the Savannah and Forest Agro-Ecological Zones in Ghana *Francis Xavier Jarawura, Nauja Kleist and Joseph Kofi Teye*	189
10	Trespassing Legal and Moral Boundaries: Ethiopian Domestic Workers Returned from the Middle East *Ninna Nyberg Sørensen*	210

PART IV The Issue of Finance

11	Climate-Related Mobility, Land and Inclusive Finance in Rural Ethiopia *Neil Webster and Adane Alemayehu Tadesse*	233
12	Conclusion: Policy Reflections on Slow-Onset Climate Mobility and the Importance of the Governance Variable *Neil Webster and Ninna Nyberg Sørensen*	253

Index   263

List of Figures and Tables

Figures

2.1	Map of Ghana showing the regions and study districts	27
2.2	Land cover changes in the Fanteakwa district between (a) January 1991 and (b) January 2020	32
2.3	Land cover changes in the Yilo Krobo district between (a) January 1991 and (b) January 2020	35
2.4	Land cover changes in the Jirapa district between (a) December 1986 and (b) December 2019	39
2.5	Land cover changes in the Wa West district between (a) December 1986 and (b) December 2019	43
3.1	The two study districts and their location in Ethiopia	57
3.2	Long-term climatic changes in Tehuledere from 1982 to 2020 in (a) average monthly temperatures (minimum, average, and maximum) and (b) average monthly rainfall	59
3.3	Long-term climatic changes in Shashemene from 1982 to 2020 in (a) average monthly temperatures (minimum, average, and maximum) and (b) average monthly rainfall	60
3.4	Rainfall trend from 1982 to 2020 (annual) and seasonal (January to February, March to May and July to September) in Tehuledere	65
3.5	Temperature trend from 1982 to 2020 (annual) and seasonal (January to February, March to June and July to October) in Shashemene	65
3.6	Temperature trend from 1982 to 2020 (annual) and seasonal (January to February, March to May and July to September) in Tehuledere	66
3.7	Rainfall anomaly from 1982 to 2021 in the Shashemene district: (a) annual; (b) January to February; (c) from March to June; (d) from July to October	68
3.8	Rainfall anomaly from 1982 to 2021 in the Tehuledere district: (a) annual; (b) January to February; (c) from March to May; (d) from July to September	69

3.9	Land cover changes in the Tehuledere district between (a) January 1986 and (b) January 2021	72
3.10	Spatial distribution of changes: (a) woody vegetation cover change; and (b) shrub/trees outside forests in the Tehuledere district	75
3.11	Land cover changes in the Shashemene district between (a) 1986 and (b) 2020	76
3.12	Spatial distribution of changes: (a) woody vegetation cover change; and (b) shrub/trees outside forests cover change from 1986 to 2020 in the Shashemene district	78
5.1	Maximum temperature time series for the Shashemene district (1983–2019)	122
5.2	Minimum temperature time series for the Shashemene district (1983–2019)	122
5.3	Normalized rainfall anomaly for the Shashemene district (1983–2019)	123
8.1	Farmland exposed to heavy erosion due to untimely rain in Burra kebele, 25 May 2019	176
8.2	Failed crop (teff) in Faji Sole kebele due to an absence of rain in the major rainy season, 21 May 2021	177
8.3	Rural-urban migrants waiting for job opportunities in one of the casual workers' sites, Shashemene, 19 May 2022	181
8.4	Rural women working as daily labourers cleaning and sorting garlic, Shashemene, 20 May 2022	182
11.1	Proportion of households with a bank account by village	241

Tables

2.1	Image datasets used for land cover change analysis	29
2.2	Satellite sensor characteristics	30
2.3	Classification of the green cover type	31
2.4	Land cover changes in the Fanteakwa district between January 1991 and January 2020	33
2.5	Land cover changes in the Yilo Krobo district between January 1991 and January 2020	37
2.6	Land cover changes in the Jirapa district between December 1986 and December 2019	41
2.7	Land cover changes in the Wa West district between December 1986 and December 2019	45
3.1	Image datasets used for land cover change analysis, Ethiopia	61
3.2	Satellite sensor characteristics, Ethiopia	61
3.3	Classification of the green cover type, Ethiopia	62

3.4	Change matrix between the tree cover and nontree cover class from time 1 to time 2	62
3.5	Change matrix between the shrub cover and nonshrub cover class from time 1 to time 2	63
3.6	Coefficient of variation of rainfall and temperature (annual and by season)	70
3.7	Inter-annual coefficient of variation, monthly rainfall contribution to the annual rainfall	71
3.8	Land cover changes in the Tehuledere district between January 1986 and 2021	73
3.9	Woody vegetation (forest/shrub land) cover change from 1986 to 2021 in the Tehuledere district	74
3.10	Shrub/trees outside forests change from 1986 to 2020 in the Tehuledere district	75
3.11	Land cover changes in the Shashemene district between January 1986 and January 2020	77
3.12	Woody vegetation (forest/shrub) cover change from 1986 to 2020 in the Shashemene district	79
3.13	Shrub/trees outside forests cover change from 1986 to 2020 in the Shashemene district	79
4.1	Determinants of the adoption of irrigation	95
4.2	Determinants of the adoption of agronomic practices (adopting improved or new crop varieties)	98
4.3	Determinants of the adoption of fertilizer application	100
4.4	Determinants of the adoption of nonfarm activities	103
4.5	Determinants of the adoption of migration strategies	105
5.1	Demographic characteristics of the respondent households	119
5.2	Perceptions of the effects of climate change by smallholder farmers (%)	120
5.3	Climate change adaptation strategies of smallholder farmers	125
5.4	Comparison of input cost and selling price of maize and teff production (2020)	127
6.1	Farmers' ownership of selected assets by kebele (%)	134
6.2	Exports of selected commodities (in US$ millions)	139
6.3	Khat growing in selected kebeles in Tehuledere (percentage of growers)	144
9.1	Perceptions of temperature	195
9.2	Perceptions of rainfall	196
9.3	Gender and climate-related forms of migration in the Wa West district (Upper West region), frequency (%)	198
9.4	Gender and climate-related forms of migration in the Yilo Krobo district (Eastern region), frequency (%)	199

11.1	Percentage of households interviewed experiencing a hazard within the last five years	239
11.2	Surveyed households' receipt of government assistance in response to drought, flooding, poor/failed harvests or lack of food, % of households	240
11.3	Accessing of credit by households, % of households interviewed	241
11.4	Proportion of surveyed households adopting a particular strategy by region	244

List of Abbreviations

1V1D	One Village-One Dam (Ghana)
ABIL	Agriculture-Based Individual Loan
CHIRPS	Climate Hazard Infrared Precipitation with Station Data
CMS	Centre for Migration Studies, University of Ghana
CRGV	Climate Resilient Green Economy
CSA	Central Statistical Agency (Ethiopia)
CSO	Civil Society Organization
CV	Coefficient Variation
DEM	Digital Elevation Model
DIIS	Danish Institute for International Studies
DN	Digital Number
ERA5-Land	The fifth generation of European Re-Analysis
ERP	Economic Recovery Programme
ESRI	Environmental Systems Research Institute
ETB	Ethiopian Birr (national currency)
FDRE	Federal Democratic Republic of Ethiopia
FGD	Focus Group Discussion
FSS	Forum for Social Studies (Ethiopia)
GAPD	Ghana Agricultural Development Programme
GBV	Gender-Based Violence
GCM	Governing Climate Mobility (research programme)
GEE	Google Earth Engine
GIS	Geographic Information Systems
GPS	Global Positioning System
GSS	Ghana Statistical Service
IFC	International Finance Cooperation
IMF	International Monetary Fund
INGO	International Nongovernmental Organization
IOM	International Organization for Migration
IPCC	Intergovernmental Panel on Climate Change
L1TP	Precision and Terrain Correction
LULC	Land Use and Land Cover

MoFA	Ministry of Foreign Affairs
NBE	National Bank of Ethiopia
NDVI	Normalised Difference Vegetation Index
NGO	Nongovernmental Organization
NIR	Near-Infrared
NSE2	Northern Savannah Ecological Zone (Ghana)
OCSSC	Oromia Credit and Saving Share Company (Ethiopia)
ODA	Overseas Development Assistance
OLI	Landsat and Operational Land Imager
PSNP	Productive Safety Net Programme (Ethiopia)
R	Red
RGB	Red, Green, Blue Bands
RMSE	Radial Root Mean Square Error
SADA	Savannah Accelerated Development Authority (Ghana)
SOCC	Slow-Onset Climate Change
SPSS	Statistical Package for Social Science (data analysis programme)
TIP	Traffic in Persons
TM	Landsat 5 Thematic Mapper
TOA	Top of Atmosphere
TOF	Trees Outside Forests
UNCDF	United Nations Capital Development Fund
USGS	US Geological Survey
WMO	World Meteorological Organization

Notes on Contributors

Adane Alemayehu Tadesse is a lecturer and researcher at the Department of Political Science and International Relations at Addis Ababa University and an affiliate researcher at the Centre for Forced Displacement and Migration Studies. He holds an MA in International Relations, a BA in political science and International Relations, and a BEd in history. He is currently a PhD candidate in political science at AAU, funded by Danida as part of the Governing Climate Mobility research programme. His research interests include hierarchies of access, climate mobility, child migration, the political economy of adaptation, differentiated vulnerability, governance and peacebuilding.

Busha Teshome is a socio-economic policy researcher with the Center for International Forestry Research. Over the past 15 years he has undertaken forest and development research in a vast range of areas including forest livelihood contribution, market structures, value chain development of forest products, cluster for forest sector development, wood-based forest enterprises (micro, small- and large-scale industries), forest cooperatives, and the climate change–migration nexus.

Dessalegn Rahmato is a Senior Research Fellow at the Forum for Social Studies, Addis Ababa. He was the Executive Director of the FSS from 1997 to 2005, and prior to that was a senior researcher at the Institute of Development Research, Addis Ababa University. He has published numerous works on land and agrarian issues, food security, rural poverty and environmental policy. His most recent book is *The Peasant and the State: Studies in Agrarian Change in Ethiopia 1950s–2000s* (2010, Addis Ababa University Press), and a short monograph entitled *Land to Investors: Large-Scale Land Transfers in Ethiopia* (2011, Forum for Social Studies).

Shelta Gatsey holds a PhD from the Department of Geography and Resource Development in the University of Ghana. She received her MA in international affairs and a BA in geography and resource development from the University of Ghana. Her doctoral thesis focuses on climate change, gender and migration in the Northern Savannah and Forest zones of Ghana and was funded by Danida as part of the GCM research programme.

Francis Xavier Jarawura is a senior lecturer in the Department of Planning at the SD Dombo University in Northern Ghana. He holds a PhD in migration studies from the University of Ghana and an MPhil degree in development geography from the University of Oslo, Norway. His key research interests include environmental change and adaptation, migration and rural livelihoods.

Kefyalew Sahle Kibret is a lecturer at Hawassa University's GIS department. He holds an MA in photogrammetry and geoinformatics and a BA in forestry. His expertise lies in GIS, remote sensing, database, land-use planning and spatial-temporal analysis. He completed the Forest Management Plan for the Addis Ababa Green Area, executed GIS and Land Use Planning projects for the Southern Nations, Nationalities, and People's Region and Sidama regions, and has published his findings in several research papers.

Nauja Kleist is a sociologist and works as a senior researcher at the Danish Institute for International Studies. Her research focuses on migration, diaspora engagement, belonging, gender, and hope and uncertainty with multisited and longitudinal studies on Ghanaian mobilities and Somali diaspora groups. Her works have been published in *International Migration Review, Ethnology, Geoforum* and *African Affairs*, and she has co-edited *Hope and Uncertainty in Contemporary African Migration* (2017).

Marie Ladekjær Gravesen is a DIIS researcher with a PhD in cultural and social anthropology from the Department of Cultural and Social Anthropology, University of Cologne, Germany. Her studies have focused on (post)colonial power relations, resource conflicts, relations between different knowledge systems and climate change adaptation. Current research interests include nature-based solutions to climate and biodiversity crises, the political economy of adaptation funds, and how conflicts over natural resources accelerate and diversify as we respond to climate and biodiversity crises.

Lily Salloum Lindegaard is a senior researcher at DIIS. She holds a PhD in international development from the University of Copenhagen, Denmark. Her work focuses on the politics and governance of climate change impacts and response from local to global scales. Her current research interests include the politics and governance of rural climate change adaptation, climate-related mobility, transformational responses to climate change, and loss and damage.

Philip Prince Kwasi Mantey is a lecturer at the School of Continuing and Distance Education, University of Ghana. He holds a PhD in development studies from the University of Ghana, an MSc in geographical information science from the University of Nottingham, and an MPhil in social change (geography) from the Norwegian University of Science and Technology. His research interests include environmental management, digital remote sensing and GIS, climate change and development education, adult education and distance learning.

Richard Seyram holds a PhD from the Centre for Migration Studies, University of Ghana. His GCM/Danida-funded thesis focuses on the relationship between governance, climate change and migration in the Upper West region of Ghana. He worked as an intern for the International Water Management Institute from November 2020 to October 2021, where he coordinated data collection on various research projects and also contributed to data analysis and report writing for the Institute.

Ninna Nyberg Sørensen is a senior researcher at DIIS. She holds a PhD from the Department of Social Anthropology, University of Copenhagen and has dedicated her career to the study of global migration dynamics, including the relationship between regular and irregular migration, multidirectional movements and its gendered forms, the workings of the migration industry, migrant disappearances and trafficking in persons. Her current work focuses on connections between climate change, mobilities and humanitarian interventions. She is attached as a visiting professor to Addis Ababa University and the Organisation for Social Science Research in Eastern and Southern Africa (2023–2027).

Joseph Kofi Teye is currently Director of Research at the Office of Research, Innovation and Development at the University of Ghana, a co-director of the UK Research and Innovation South-South Migration, Inequality and Development Hub, and the immediate past Director of the Centre for Migration Studies, University of Ghana. He holds a PhD in geography from the University of Leeds and an MPhil in social change from the Norwegian University of Science and Technology. His current research interests include migration and development, migration policy development, environmental change and natural resource governance.

Neil Webster was a senior researcher at DIIS (2003–2023) and Principal Investigator for the GCM Research programme. He holds a PhD and an MA in sociology from Manchester University, an MSc in International Relations from the London School of Economics, and a BA in economics and sociology from Durham University. He has researched and published extensively on governance and local development issues with a focus on South Asia. From 2009 to 2014, he worked for the United Nations Development Programme and Capital Development Fund as senior local governance advisor in Nepal, and regional local governance adviser for the UNCDF in Africa.

Zerihun Mohammed is Executive Director of Good Governance Africa (Eastern Africa) and was formerly a senior researcher at the Forum for Social Studies, Ethiopia. He holds a PhD from the University of Cambridge in human geography. He has extensive research experience in subjects including migration, social relations, livelihoods and resource management.

Acknowledgements

This book is the outcome of a joint research endeavour aimed at understanding the complex relationship between slow-onset climate change, human mobility and governance in two African countries, Ethiopia and Ghana. Both countries are experiencing environmental change that is directly affecting livelihoods and mobility, but in which the governance context varies considerably. The bulk of the empirical data collection, the methodological tools and the theoretical approach(es) were developed as part of a four-year interdisciplinary research programme entitled 'Governing Climate Mobility' (GCM). The GCM Programme involved 15 Ethiopian, Ghanaian and Denmark-based scholars and was carried out over a four-year period (2019–2023). Writing continued throughout 2024.

A particular debt of gratitude is owed to the GCM researchers in Ghana and Ethiopia for their dedication and hard work in seeing the project through in the face of serious and unforeseeable challenges that included the COVID-19 pandemic, the armed conflicts between Ethiopia's Federal Government and the Tigray People's Liberation Front (2020–2022) and Amhara Militias (throughout 2023), and the revolving conflict in Northern Ghana. The editors of this volume and GCM programme management are also grateful to the enumerators and field assistants, to the many government officials at the national and local levels who contributed viewpoints and analysis along the way, to the administrators, finance officers, logistical supporters and drivers from the three collaborating research institutions: the Danish Institute for International Studies, the Forum for Social Studies in Addis Ababa, and the University of Ghana. Their support has been invaluable.

Second, we wish to express our gratitude to the Danish International Development Agency (Danida) and Denmark's Consultative Research Committee for Development Research (FFU), which generously supported the research programme and also provided an extra grant that enabled us to organize a final workshop in Addis Ababa in February 2023, during which the chapters included in the volume were critically discussed and feedback was given to the authors from peers in the wider research, government and nongovernmental organization communities. Thank you to the student assistants at DIIS, Frederikke Møller Helvad and Sofie Amalie Poulsen, for assisting in finalizing the manuscript for publication.

Last, but most importantly, we would like to express our deep appreciation to the many women and men in Ethiopia and Ghana who gave their time and offered valuable insights along the way. Without their participation, there would be no book. We therefore dedicate the various chapters to them and to future generations living in climate change affected localities in Ghana and Ethiopia.

1

Introduction: An Outline of the Analytical Approach to 'Governing Climate Mobility'

Ninna Nyberg Sørensen, Lily Salloum Lindegaard, Neil Webster, Joseph Kofi Teye and Zerihun Mohammed

Introduction

As climate change progressively affects the physical environment, undermines livelihoods and threatens human security, understanding the broader impacts represents a serious challenge for the 21st century. We know that climate change shapes socioeconomic dynamics in ways that intensify household vulnerability and insecurity (Ribot, 2014; Stapleton et al, 2017; Gebre and Rahut, 2021; Mthethwa and Wale, 2023). It is also established that environmental degradation affects human mobilities both directly and indirectly (Borderon et al, 2019; Hub, 2020), interacts with several other drivers of migration, and hits the most exposed hardest (Foresight, 2011; Afifi et al, 2014; Crawley et al, 2022), including migrants as well as those too poor, unable or unwilling to move (Bettini, 2017; Zickgraf, 2021; Boas et al, 2022). A review of contemporary literature on human mobility will find that migration has the potential to reduce inequalities. Yet not everyone is able to move or, if they do so, is able to reap the potential benefits of migration (Bettini and Gioli, 2016). Therefore, despite an abundance of literature on climate change and migration respectively, the nature of this relationship remains subject to contestation.

This is surprising given, on the one hand, that the number of studies on the relationship between climate change and human mobility is rapidly growing and, on the other hand, that the relationship between the environment and human mobility is not a new area of discovery within migration research. Environmental factors played an important part in early (late 19th-century) migration theorizing, in which the search for better land, milder climate conditions and better livelihood options were given importance in explaining the reasons behind population movements, what today would be considered a form of adaptation to climate change. Over the course of the 20th century, environmental considerations strangely ceased to be given importance – according to Piguet (2013), due to a combination of the idea that progress implies a decreasing importance of natural factors, a vigorous rejection of (natural) determinism, the rise of economic push-pull factors in migration theory and the constitution of a specific field of refugee/forced migration studies around a rights-based political paradigm. This lasted until alarmist projections presented 'environmental refugees' as one of the foremost crises of our times in the early 1990s, projections that later became subject to critique for lacking methodological rigour (Bettini, 2012; Jacobeth and Methmann, 2012; Piguet, 2013).

Over the past decade, climate change and mobility dynamics have become important focus areas in both policy and academic circles. Empirical evidence has been gathered, yielding insights about the variety of historical and contemporary patterns and contexts of the phenomena. Climate change literature typically distinguishes two broad categories of change according to the speed with which the change occurs, namely sudden or rapid-onset events such as floods or storms, and slow-onset events such as rising temperatures and drought. The present volume focuses on slow-onset events and thereby the long-term trends in climate change, which despite immensely harmful effects have received less attention than sudden-onset climatic shocks and the related disaster risk management strategies (Zickgraf, 2021). The literature also discusses whether migration could and should be considered an adaptation strategy to climate change, the arguments alternating between the role migrant remittances may play in financing local adaptation and viewing migration as maladaptation (Warner, 2010). The mainstreaming of development and governed migration as a conducive contribution to adaptation is already adopted by powerful international organizations, although often without addressing the political rather than the environmental factors affecting climate change and human mobility (Bettini and Gioli, 2016). To address this neglect, the volume explores the possible links between slow-onset climate change, mobility and adaption, in particular the ways in which governance contexts and interventions may play into such links.

This volume posits that attention to the politics and governance of climate mobility is more important than is often realized for understanding environmental impacts and mobility responses. Diverse governance actors shape how climate change is experienced and by whom, and also mediate people's responses (Ionesca et al, 2017; Piguet, 2021). It clearly emerges across the volume's chapters that local land and resource management affect how climate change impacts on agrarian livelihoods; international mobility regimes affect migration preferences and routes; and national adaptation policies and programming affect local climate impacts and responses. More informal forms of governance – enacted by traditional and religious leaders, local elders and community groups – are also significant in shaping local conditions and norms, as are aspirations and opportunities for those who move or those who stay behind. Overlooking such layers of governance – approached here non-normatively as 'the activity of governing' (Rose and Miller, 1992; Bulkeley and Stripple, 2013) – will limit understandings of environmental impacts and mobility responses, and thereby scholarship, policy and practice. Pushing further, we join recent assessments of the overlooked politics of climate-related mobility (Boas et al, 2022). In doing so, we argue for an approach that not only identifies particular forms of governance and their effects, but also seeks to document how these have come to be over time and, in the context of increasing climate impacts and diverse mobility responses, how these might be reconfigured and changed in response to changing circumstances. This, we believe, fills an important gap in the climate-related mobility literature.

This volume sets out to address these needs through analyses of two African countries: Ethiopia and Ghana. These are both countries where environmental degradation is directly affecting livelihoods and mobility but where the governance context varies considerably. There are significant differences in the roles of local government and the policies they implement, in the roles of informal institutions and traditional authorities, in the politics of each state's formation, in their democratic conditions and in their strategies for economic growth. For the purposes of our analysis, the differences between these two country contexts and the multiple research sites within each make it possible to identify configurations and contributions of characteristics across different localities. This allows us to identify broader tendencies relevant to cases beyond the selected countries and to climate mobility policy and politics on a wider scale. Based on subnational field studies in rural and peri-urban areas of Ethiopia and Ghana, the chapters bring agricultural development and urbanization in contemporary Africa into the discussion, contributing to the broader exploration of climate mobility practices in challenging contexts.

Climate change, migration and governance tend to be studied as separate fields of research, resulting in their theoretical contributions, conceptual frameworks and methodologies not always sitting easily together. This volume builds on elements from these separate literatures. With a central objective being to better understand how different governance contexts affect adaptive climate mobility, it breaks with literature dominated by direct causal attributions of environmental factors to migration (for further critique of direct causality, see Piguet, 2013; Adger et al, 2015; Zetter, 2017; Ionesca et al, 2017). Instead, the impact of climate change on households and local communities is viewed through the lens of governance, both formal and informal, as it is exercised across scales and experienced on the ground. The mobility of households and individuals in contexts affected by climate change is explored through personal experiences of climate change, which in turn are understood as being mediated by the practices of formal and informal institutions. These are then viewed through the lens of national government strategies and policies and the political and administrative nature of their impact locally, not least in influencing the practices and parameters of the institutions active in local governance. As these are often influenced by the views, interests and policy imperatives of the Global North (Crawley et al, 2022), the processes through which this manifests itself are also analysed.

With these aims as our point of departure, this introduction intends to unfold diverse analytical perspectives and approaches to the role of governance in mobility responses regarding slow-onset climate change. The intention is not to provide an exhaustive or overarching theory, but rather to introduce concrete attempts at theorizing governance in debates on climate-related mobilities and, through this, to promote discussion and further scholarly engagement in this topic.

From climate displacement to climate mobilities

Shifting the focus from climate-related displacement and migration to mobilities more broadly directs attention to (im)mobility hierarchies, power dynamics and differentiated contexts shaping mobility processes. It makes it possible to analyse climate-related movements through a broader lens, accounting for various forms of (im)mobilities that emerge in response to climate change, taking note of the sociopolitical contexts shaping people's aspirations and capacities to mobilize their existence. The theoretical developments underlying this approach will be addressed later on.

The field of climate mobilities has grown out of 'environmental studies' that developed in relative isolation from the general migration theory literature. Early environmental migration studies often relied on predictive science, resulting in disproportionate power being awarded to model-based understandings of social life and mobility (Hunter and Simon, 2023). The somewhat simplistic and geodeterministic discourses on 'environmental refugees' or 'climate refugees' grew out of this tradition, but also gave rise to debates around the conceptual ambiguities relating to categorization. As no legal definition of a 'climate refugee' exists, the concept of 'climate migrant' gained traction, but was found to be just as limiting in its scope. Currently the relationship between climate change and human mobility is seen as being complex and multifaceted, going beyond the voluntary/forced and proactive/reactive divides (Hugo, 2011; Hunter et al, 2015; Ferris,

2020). This complexity finds its expression in the widely acknowledged heuristic framework on migration drivers (Black et al, 2011), which points out that migration in the context of climate change is driven not only by environmental factors but also by the interaction with existing economic, political, social and demographic drivers on the macro-, meso- and micro-levels. With drivers understood as forces leading to the inception of migration and the perpetuation of movement, Van Hear, Bakewell and Long (2018) have further suggested to distinguish between predisposing, proximate, precipitating and mediating drivers, each having different potential for political intervention.

The literature on drivers has been particularly helpful in empirical case studies directed at the complex response chains of individuals or their families and households when faced with the need to leave or stay, as well as the diversity of ways to do so (for example, Borderon et al, 2019; Cattaneo et al, 2019; Hoffmann et al, 2020). Such studies provide fresh empirical evidence that climate and environmental changes are factors adding to and/or impacting on conventional drivers of migration (Bettini and Gioli, 2016).

Other conceptual frameworks and theoretical reflections have been useful in describing the different forms of mobility and immobility to which climate change has given rise. The aspiration-(cap)ability framework developed by Carling and Schewel (2018) and de Haas (2021) allows for a better understanding of human mobility as an intrinsic part of broader processes of social and environmental change, as well as of the relationship between structure and agency within concrete situations. By breaking up migration outcomes in two separate steps – on the one hand, the evaluation of migration as a potential course of action in a given context and, on the other hand, the realization of mobility or immobility at a given moment, this approach introduces less deterministic ways of comprehending mobility outcomes of climate change impacts.

The application of the literature on mobilities (Sheller and Urry, 2006; Cresswell, 2010) and mobilities in in the context of climate change (Wiegel et al, 2019; Zickgraf, 2021; Boas et al, 2022; de Sherbinin et al, 2022) also allows for the inclusion of diverse forms of movement: circular forms of mobility, such as seasonal mobility typical in agrarian settings, onwards-migration from an initial destination, displacement, resettlement, return movement and forced immobility, among others (McLeman and Hunter, 2010; Chen et al, 2017; Rigaud et al, 2018). Moreover, attention to complex trajectories (Schapendonk et al, 2014) – including instances in which mobility practices are episodic, seasonal or multidirectional (progressive with intermediate destinations along the way rather than between two points with an ultimate destination) – are recent developments that offer an analytical lens for a better understanding of the complexity and diversity of the relationships between climate change and mobility, and the human responses to environmental stress. So are considerations of the translocality of households formed both by members who have migrated and others who have remained in their places of origin (Rockenbauch et al, 2019; Porst and Sakdapolrak, 2020), the latter's ability to stay in place often being enabled by the former.

The literature referred to previously has underlined that the diversity of migration potentials linked to environmental change presents challenges to institutions and policies that are not designed to cope with the impacts of complex causality, surprises and uncertainty about socioecological thresholds (Warner, 2010). As stated by Piguet, the lack of governance and power analysis is 'the most striking weakness of the current research on environmental migration', where '[i]t remains difficult to contextualise empirical results within larger sets of power relations and governance' (Piguet, 2021: 10). Similar critiques are raised by Boas et al (2018, 2022), Geddes et al (2012) and McCarney and Kent (2020). Thus, beyond overall policies to limit environmental degradation, it remains a challenge to develop mobility policies that do not simply attempt to enable or force people to either stay in place or migrate (Lavenex and Piper, 2021). However, for now, the study of governance is often siloed in separate studies of climate or migration governance, while empirical case studies of climate-related mobility rarely examine the political factors at play (Zickgraf, 2021). Engaging the two in a common analytical framework will be one of the central challenges over the next ten years (Piguet, 2021).

In climate governance studies, Methmann and Oels (2015) have traced the genealogy of environmental migration from a 'pathology to be prevented' over an issue of refugee rights to becoming a 'rational strategy of adaptation' to unavoidable levels of climate change (Methmann and Oels, 2015: 51). Thus, while mobility – under certain circumstances and for some groups – offers an effective form of adaptation to the adverse effects of climate change, it may increase impoverishment and deepen vulnerability for others (Vinke et al, 2020). Whether migration advances adaptation or produces precarity is therefore an empirical question to be studied, always considering migrant positionality and the ways in which migrants' social practices are structured by dimensions of social inequality, including gender, class, age, ethnicity and regional origin (Porst and Sakdapolrak, 2018). At the same time, the derivative shift from governing climate change as risk management to building resilience tends to occlude the political, as responsibility for resilience is placed on the potential victims of the effects of climate change. It is the contention of this volume that a shift in focus from (reactive) displacement or (proactive) migration to a non-normative examination of mobility responses more broadly allows for a more fine-grained analysis of the mobility hierarchies, power dynamics and differentiated meanings that shape any mobility process.

Unfolding the governance perspective

This volume's governance perspective supports a non-normative examination of the role of diverse forms of governance in shaping mobility contexts and responses. Governance and politics have long been implicit in studies of climate-related mobility. As outlined earlier, early discourses on 'climate refugees' and 'adaptive climate migrants' were normatively charged and linked to policy agendas, from securitization (Brzoska and Fröhlich, 2016) and protection agendas (Gemenne, 2015) to developmentalist agendas (Bettini and Gioli, 2016). Thus, existing literature on climate-related (im)mobilities is tied to particular normative and policy framings, interests and aims. Increasingly, scholars are explicitly interrogating such aspects of governance, politics and power as part of non-normative, analytical examinations of climate mobilities (Boas et al, 2022).

To help break new ground on the emerging strand of climate mobility literature, this volume presents an array of approaches to understanding how governance and politics shape mobility contexts and responses in localities affected by climate change. Woven through the chapters is the understanding of forms of 'political authority' exercised by the diverse institutions that govern territories, people and resources (Lund, 2006), which shape climate affected contexts and mobility responses. This broad conceptualization provides room for situated analyses exploring a variety of empirical contexts, in line with our grounded empirical approach to studying governance and climate mobility. This approach also allows the individual contributions room to explore vastly different constellations and aspects of political authority, as well as relations between political authority, climate-affected contexts and mobility practices.

In considering the role of political authority, we seek to bridge siloes between scholarly fields of mobility, governance, and environmental and climate change. To do so, we identify points of coherence that emerge across the literature and are indeed present as cross-cutting themes within this volume. These include plurality in governance actors and practices, scalar dynamics of governance, and differentiated impacts and responses – as well as input to governance itself. These themes provide useful points of departure for new analytical lines of enquiry that can respond to elements of governance and climate-related mobility emerging in our own grounded research and beyond. They relate to questions around how governance is exercised and experienced and with what effects for whom. We present these themes briefly here, as they emerge across literatures on mobility, governance, and climate and environmental change.

Governance emerges as the effect of combined, sometimes contradictory effects of a plurality of actors (Lund, 2006: 686; see also Rose and Miller, 1992: 176). While some analyses of governance focus on formal state-based institutions – for example, ministries, national and district assemblies, municipal governments and international fora – informal actors such as religious, traditional or nongovernmental authorities have also received attention (Lund, 2006) and are highly relevant for understanding climate affected contexts and mobility responses. Formal state institutions shape, among other things, vulnerability and adaptation (Ribot, 2014; Nightingale et al, 2017) with potential implications for climate-related mobility (Warner, 2010; Clement et al, 2021). This is the case, for instance, through resource management, livelihood support, infrastructure interventions, social services and so on in climate change-affected areas. Formal actors can also shape climate-related mobility more directly, for example, regarding internal migration (Rigaud et al, 2018; Clement et al, 2021) or climate resettlements (Arnall, 2019; Lindegaard, 2020a). Beyond the state, other formalized actors, including international or nongovernmental organizations, also shape many of these factors, from aspects of livelihoods to mobility. In addition, informal governance actors such as community groups, traditional authorities, and migration networks and associations are often also highly relevant in people's livelihood opportunities, resource access and mobility patterns (Sikor and Lund, 2009; Sørensen, 2016; Seter et al, 2018) – even in terms of service provision in some instances (Rohregger et al, 2021).

This understanding of political authority as the aggregate of a pluralism of actors and institutions has gained ground over the last few decades (Abrams, 1988 [1977]; Mitchell, 1991, Rose and Miller, 1992; Lund, 2006; Bierschenk and Olivier de Sardan, 2014). Now, it is emerging in literature on climate mobilities, for instance, regarding 'climate mobility regimes', understood as the 'interconnected sets of socio-economic and political relations consisting of different types of actors, that frame, manage, and regulate the nexus between mobilities and climate change ... resulting in particular modes of governing of climate mobilities' (Boas et al, 2022: 3371). Governance in relation to climate-related mobility can thus be usefully approached as a complex plurality of actors and institutions that shape climate-affected contexts, differentiated impacts and possible responses, including mobility, in a variety of ways. For scholars, as well as practitioners and policy makers, this indicates the importance of attention to a variety of both formal and informal actors when seeking to understand the governance context shaping climate-affected contexts and mobility responses.

An important aspect within governance plurality is that of scale – an alternative to conceptualizations of fixed hierarchies of levels (for example, international, national or local; see Brenner, 2001: 591–592) – that instead emphasizes relations in nonconcentric constellations. This can include the role that international organizations or donors play in shaping local contexts and mobility responses, the intrascalar and translocal connections through migrant networks, or the way in which local communities might engage with and lobby for climate responses from actors at a national level – all of these being examples emerging in the contributions to this volume. The concept of scale has also become influential in environmental and climate change adaptation literature, for instance, regarding how international actors shape local contexts (Urwin and Jordan, 2008; Adger et al, 2015; Lindegaard, 2020b). In the mobilities literature, scale is an important constitutive element in the scaled domestic, international and regional dynamics that shape mobilities, and a host of related relations, within and across localities (Sheller and Urry, 2006; Lavanex and Piper, 2021). Empirically, scalar dynamics can be seen in terms of how migration networks and hometown associations facilitate translocal flows of resources, information and migration itself (Kleist, 2014; Sørensen, 2016), or in the cross-scalar relationship between mobilities and immobilities (Weigel et al, 2019). As is also indicated in many of the contributions in this volume, governance actors from multiple scales are often simultaneously at play in shaping climate-related mobility contexts and practices. This volume thus underlines how multiscalar governance – of intersecting governance roles and regimes across and within scales – is a constitutive aspect of the spaces in which climate-related mobility is practised, as well as of mobility options and opportunities themselves.

Finally, mobility, governance and climate and environmental change literatures all emphasize various aspects of differentiation across demographic or social groups – in essence, how some groups of people or individuals with particular age, gender, religious or ethnic group characteristics might experience climate change differently, have varying opportunities to influence climate or mobility governance, or have different options for adaptation or mobility. Vulnerability and political ecology literature, for instance, documents how historical and current interventions and relations on development, marginalization, access, extraction and so on contribute to differentiated climate change impacts within particular contexts (Adger et al, 2009; Ribot, 2014; Taylor, 2014). The mobility literature, in turn, documents how vulnerable and marginalized populations are more often immobile, illustrating the spatial injustices and uneven mobilities linked to climate change (Sheller, 2018; Weigel et al, 2019). This literature specifically highlights the 'differentiality of im/mobilities, or why people have uneven capabilities and aspirations for im/mobility practices, grounded in both personal and structural factors' (Weigel et al, 2019: 4), thus underscoring the contextual factors shaping mobility as an option, such as that similar forms of climate-related mobility can increase vulnerability for some, while reducing it for others (Vinke et al, 2020). Related findings have emerged in adaptation and development literatures, which find that interventions seeking to reduce climate change impacts can also (and sometimes simultaneously) exacerbate vulnerability and lead to maladaptation or reproduce vulnerability (Eriksen et al, 2021), again occurring unevenly within a population, linked to existing vulnerabilities or inequalities. Authors also describe how such outcomes, and the governance dynamics producing them, are tightly bound to political and power dynamics (Eriksen et al, 2015; Nightingale, 2017).

The cross-cutting themes of scale, diversity in governance actors and differentiated impacts help characterize governance in relation to climate-related mobility and consider its effects. Yet the question of how particular governance configurations come to be, and how these might be changed over time, remains. In emphasizing these questions, we seek to further recent scholarship on the significance of politics and power in shaping climate-related mobility (Boas et al, 2022). We simultaneously argue for an approach that not only identifies forms of governance and their effects but also interrogates how these have come to be over time and how these might be reconfigured to better address increasing climate impacts and diverse mobility responses. We build on the literature on political authority that emphasizes how authority is fluid and constantly (re)produced and challenged in a variety of ways (see, for example, Nightingale, 2018, Nightingale et al, 2022). Rather than being independent from societies and environments, governance emerges through interactions with them. Individuals and groups as well as environments thus influence governance in a process of co-production (Lund, 2016; Christoplos et al, 2017; Siddiqi and Blackburn, 2022). Such approaches have become influential in relation to climate change adaptation, where authors describe how climate-related interventions, and the ways in which they reshape local realities and existing responses, are themselves political arenas with elements of contestation and bottom-up change (Eriksen et al, 2015; Lindegaard and Sen, 2022). This indicates that governance of climate-related mobility should be understood not as a one-way process of governance over climate-related mobility, but as multidirectional. It points to the centrality of a structure-agency interplay in informing governance of climate-related mobility (see Boas et al, 2018; Webster, 2023), where structures of authority are (re)produced through their interactions with populations in climate-affected contexts. Such an approach is fundamentally at odds with simplistic conceptualizations of climate refugees bound by structural constraints and lacking agency, or climate migrants who are seen to exert agency despite such constraints. It thus provides an alternate starting point for understanding structure and agency within climate-related mobility and is also essential for studies of governance of climate-related mobilities, which otherwise could easily overemphasize structure or undertheorize agency.

To summarize, we argue that the most fruitful examinations of climate-related mobility will be situated in the interaction between agency and structure, with an eye to political authority and politics. Together, these analytical entry points regarding governance – plurality, scale and differentiation – can help us better understand the diverse forms that climate mobility governance takes empirically. This will help shed light on how climate mobility emerges in particular forms in particular contexts, why some groups or individuals are more likely to stay or move in certain ways than others, what actors are involved and how, and how this might change over time. Further, attention to underlying political dynamics is a necessity for analysing processes of and potentials for (re)production and contestation of existing governance approaches. This can support further research as well as policy and programming efforts to support adaptive forms of climate-related mobility.

Collaborative research

The research presented in this volume stems from a collaborative research programme carried out jointly by the Centre for Migration Studies (CMS) at the University of Ghana, the Forum for Social Studies (FSS) in Ethiopia and the Danish Institute for International Studies (DIIS). Over its lifetime from March 2019 to December 2023, the programme has been generously funded by the Danish Ministry of Foreign Affairs through its collaborative research, knowledge exchange and capacity-building funding scheme, administered by the Danida Fellowship Centre. The funding has enabled senior researchers at the three participating research institutions to dedicate a substantial amount of their time to developing and conducting the research underlying the findings included in the volume. It has also enabled to fully fund three PhD projects. These projects have added theoretical, methodological and empirical richness to the overall programme and resulted in several co-authored chapters across nationality and academic titles.

The contributions

The chapters included in this volume provide complementary analyses of the effects of climate change in Ethiopia and Ghana, whether and how households and individuals engage in mobility to adapt to climate change, and the extent to which climate-related and mobility governance processes at the national and local levels are conducive to such practices. These key questions are addressed by looking at the broader politics of climate adaptation and its implications for climate mobility options.

The volume consists of 12 original chapters, organized into four parts. The two chapters that make up Part I support the rest of the volume by providing an assessment of climate variability and vegetation cover in the study districts over the past 35 years. Both chapters employ Geographic Information Systems (GIS) techniques to identify and quantify observable rates and changes which they then triangulate with survey and interview material collected at the national and district levels.

Chapter 2, 'Vegetation Cover Changes in the Eastern and Upper West Regions of Ghana' by Teye, Mantey, Gravesen, Richard, Lindegaard, Jarawura, Kleist and Gatsey, detects changes over the period from 1986 to 2020. The authors demonstrate that the natural vegetation is of great importance to regional hydrology, carbon storage and global climatic processes. It is also of great economic value. However, both the Eastern and Upper West regions have experienced rapid vegetation loss due to multiple causes, including the clearing of land for settlement, logging, mining and agricultural activities, but also demographic pressure, poverty and inadequate government policies, all impacted by climate change with negative effects on local livelihoods.

Chapter 3, 'Land Cover and Land Use Change and Climate Variability: Evidence from Longitudinal Geographic Information System Data in Shashemene and Tehuledere' by Kefyalew, Zerihun, Lindegaard, Sørensen, Webster and Dessalegn, applies a similar approach to reveal considerable changes in shrub and tree cover, particularly in agricultural and forested areas. Interestingly, an increase in trees in settlement areas and along streams, where eucalyptus is becoming prevalent and important to households' livelihood strategies, is also detected. This is a new development in rural livelihoods, which are mutually shaped by governance contexts and household strategies, with implications for the local environment.

Both chapters point to the need to take account of multiple causations when addressing climate change impacts and make evident that attention to autonomous adaptation strategies is paramount for the development of sustainable resource management policies.

Part II focuses on observable local impacts and adaptation strategies in particular places. Special attention is given to the perception of changing weather conditions among rural and semi-urban populations and how they have coped with this over time.

In Chapter 4, 'Assessing Adaptive Capacity to Climate Change and Variability in the Savannah and Forest Agro-Ecological Zones of Ghana', Jarawura, Teye, Lindegaard and Kleist explore adaptive capacities through a comparative and cross-scalar approach that focuses on the interaction between households' own preferences and strategies, and the institutions and governance contexts that shape available options. They find that location in an acro-ecological zone and household characteristics determine the capacity to adopt adaptation strategies. Households in the forest zone have a unique location in that they experience two rainy seasons, enabling crop cultivation twice a year, spreading risk and increasing incomes. In contrast, households in the Savannah zone experience only one rainy season, leaving them more vulnerable to the vagaries of the weather and changes in the climate. Limited access to environmental and developmental resources explains why southward migration in search of jobs is an adaptation strategy often adopted here.

In Chapter 5, 'Climate Change and Its Implications for Smallholders' Crop Production: Change from Maize to Teff to Haricot Beans in the Shashemene District, Ethiopia', Busha and Zerihun assess local-level adaptation practices among smallholders depending on crop farming. Due to unpredictable rainfall and temperature changes, crop production is severely affected. To maintain their livelihoods in unfavourable climatic conditions, smallholders have applied several adaptation strategies, including improved or new varieties of crops, increased use of fertilizers and pesticides, engagement in non-agricultural work and seasonal migration. More than half of the households are nevertheless forced to reduce their expenses, which typically means reducing the size and number of meals. The study points to the need to support smallholder farmers in coping with existing climatic conditions and to adapt for future climate change in ways that reflects their agro-ecological and socioeconomic context.

In Chapter 6, 'Khat Cultivation and Climate Change in Tehuledere, South Wollo', Dessalegn explores how khat cultivation has become integrated in local rural livelihoods over the past few decades. Khat, which in the past was associated with faith-based customary practices and its cultivation confined to a small area in the southeast of Ethiopia, has spread to rural areas across the country. The crop's main attraction is its high market value, the relatively small amount of land involved, and the fact that it can be harvested three times in a year, providing a steady cash income. The shift to khat cultivation alongside other crops expands households' food security and acts as an adaptation strategy to climate change. The chapter reveals how khat cultivation has been integrated into the dynamics of many rural livelihoods, affecting family and gender relations, the labour process, asset management, community relations and migration, as well as relations between khat growers and state actors.

Policy that seeks to limit climate-related migration through adaptation efforts is increasingly implemented throughout the Global South. In Chapter 7, '"They Are Supposed to Stay Home": Examining the Politics of Adaptation and (Im)Mobility in Northern Ghana', Lindegaard, Jarawaru and Kleist examine such efforts in northern Ghana where irrigation schemes linked to the 'One-Village – One Dam' initiative aim at curbing out-migration in areas adversely affected by climate change. The chapter scrutinizes the politics and outcomes of the initiative, emphasizing the differentiated perspectives among actors within and outside two affected communities. The analysis shows that competing portrayals of climate-related mobility are produced, mobilized and contested in social and political spheres by a variety of actors for different purposes, and questions efforts too narrowly directed at limiting mobility through in-situ adaptation projects. Approaches that support the choice of affected populations to leave or stay may in fact provide better adaptive outcomes.

Chapter 8, 'Short-Distance Migration as an Adaptation Strategy to the Impacts of Climate Change: The Case of Shashemene District', provides fuel for this argument. In the Shashemene district of Ethiopia, Zerihun's analysis shows how smallholder farmers employ both short-term and long-term migration to cope with challenges posed by environmental change. Variation in distance to destination and duration of stay is closely related to financial, natural, physical and human resources. Short-term migration to nearby urban centres is often the only available option for those who lack the financial resources to go abroad. The financial gains are limited. Women and young girls' engagement in casual work and petty trade is nevertheless breaking down old, gendered taboos, enabling women to become the main provider for their families. The downside to this is higher school dropout rates, as the eldest (often the girls) must stay at home and look after younger siblings while their mothers work in town.

The two chapters that make up Part III make gender the focal point of the analysis. In Chapter 9, 'Gender Dimensions of Climate Change-Related Migration in the Savannah and Agro-Ecological Zones in Ghana', Jarawura, Kleist and Teye focus on the cultural differences between the two Ghanaian study sites and, more particularly, the influence that the patriarchal inheritance system in the Savannah and the mixture of patriarchal and matrilinear inheritance systems in the forest area may have on mobility practices and responses to climate change. In both regions, people identify climate change as an important driver of migration and point to both intensification and diversification of migration patterns as a response. Men dominate the climate-related migration streams in both areas, but women increasingly participate. The lower participation rate among women in the Savannah area is attributed to the patriarchal control of women's sexuality, as women migrating on their own become subject to rumours of having engaged in sex work.

How women's migration is understood and categorized is also the focus of Chapter 10, 'Trespassing Legal and Moral Boundaries: Ethiopian Domestic Workers Returned from the Middle East', in which Sørensen demonstrates how gender affects the way in which migration takes place, the risks faced prior to, during and after migration, and the employment and protection options available. Humanitarian organizations predict that the combined effects of environmental change, economic hardship and protracted conflict situations will increase Ethiopian women's vulnerability to human trafficking. This legitimizes European pressure on African governments to 'combat' irregular migration. A derivative effect of the coinciding focus on vulnerability and control is that women's irregular migration routinely becomes conflated with human trafficking. The chapter suggests reorienting the analytical focus away from questions of (il)legality and instead taking women's lives and expose to danger at home and abroad into account.

Over the years, development agencies have underscored the importance of 'inclusive finance', but have focused their attention mostly on enhancing local economic initiatives and activities. Part IV takes up this issue. In Chapter 11, 'Climate-Related Mobility, Land and Inclusive Finance in Rural Ethiopia', Webster and Adane argue that slow-onset climate change is often provided as a causal explanation for domestic as well as international mobility. However, while slow-onset climate change is an important factor, the basis for the mobility decisions taken by rural households – particularly smallholders with a strong element of subsistence farming – varies greatly. Factors influencing these decisions include environmental conditions, but also the presence or absence of state provision in areas such as infrastructure and social protection, and not least the nature and regulation of land rights. The specific characteristics of a household also play a key role. The potential for many smallholder farmers to engage in commercial agriculture is considerable. Enabling more accessible finance could shape the space in which decisions on mobility are being taken. Mobility would undoubtedly remain with a successful inclusive finance strategy, but its form and role could be changed significantly, not least contributing to a more transformative approach in Ethiopia's agrarian economy.

By way of a conclusion, Chapter 12 by Webster and Sørensen, 'Conclusion: Policy Reflections on Slow-Onset Climate Mobility and the Importance of the Governance Variable', draws together the main findings and makes suggestions for future policy engagement.

Together, the chapters make a nuanced and grounded representation of climate change impacts and (im)mobility responses in Ghana and Ethiopia. It is beyond doubt that climate change progressively threatens human security, undermines livelihoods and affects established mobility patterns in the selected study sites. Yet, no linear or direct causal relationships can be established. What can be established is that varying governance contexts at multiple scales and levels shape such relationships, sometimes with unintended, but severe consequences. The multiscalar analysis applied throughout the chapters highlights the underlying inequality dynamics through which adaptive capacity and mobility strategies are governed.

References

Abrams, P. (1988 [1977]) 'Notes on the difficulty of studying the state', *Journal of Historical Sociology*, 1: 58–89.

Adger, W.N., Dessai, S., Goulden, M., Hulme, M., Lorenzoni, I., Nelson, D.R., Naess, L.O., Wolf, J. and Wreford, A. (2009) 'Are there social limits to adaptation to climate change?', *Climatic Change*, 93: 335–354.

Adger, W.N., Arnell, N.W., Black, R., Dercon, S., Geddes, A. and Thomas, D.S. (2015) 'Focus on environmental risks and migration: causes and consequences', *Environmental Research Letters*, 10(6): 060201.

Afifi, T., Liwenga, E. and Kwezi, L. (2014) 'Rainfall-induced crop failure, food insecurity and outmigration in Same-Kilimanjaro, Tanzania', *Climate and Development*, 6(1): 53–60.

Arnall, A. (2019) 'Resettlement as climate change adaptation: what can be learned from state-led relocation in rural Africa and Asia?', *Climate and Development*, 11(3): 253–263.

Bettini, G. (2012) 'Climate barbarians at the gate? A critique of apocalyptic narratives on "climate refugees"', *Geoforum*, 45: 63–72.

Bettini, G. (2017) 'Where next? Climate change, migration, and the (bio) politics of adaptation', *Global Policy*, 8: 33–39.

Bettini, G. and Gioli, G. (2016) 'Waltz with development: insights on the developmentalization of climate-induced migration', *Migration and Development*, 5(2): 171–189.

Bierschenk, T. and Olivier de Sardan, J.P. (2014) *States at Work: Dynamics of African Bureaucracies*. Leiden: Brill.

Black, R., Adger, W.N., Arnell, N.W., Dercon, S., Geddes, A. and Thomas, D. (2011) 'The effect of environmental change on human migration', *Global Environmental Change*, 21: S3–S11.

Boas, I., Kloppenburg, S., van Leeuwen, J. and Lamers, M. (2018) 'Environmental mobilities: an alternative lens to global environmental governance', *Global Environmental Politics*, 18(4): 107–126.

Boas, I., Wiegel, H., Farbotko, C., Warner J. and Sheller, M. (2022) 'Climate mobilities: migration, im/mobilities and mobility regimes in a changing climate', *Journal of Ethnic and Migration Studies*, 48(14): 3365–3379.

Borderon, M., Sakdapolrak, P., Muttarak, R., Kebede, E., Pagogna, R. and Sporer, E. (2019) 'Migration influenced by environmental change in Africa', *Demographic Research*, 41: 491–544.

Brenner, N. (2001) 'The limits to scale? Methodological reflections on scalar structuration', *Progress in Human Geography*, 25: 591–614.

Brzoska, M. and Fröhlich, C. (2016) 'Climate change, migration and violent conflict: vulnerabilities, pathways and adaptation strategies', *Migration and Development*, 5(2): 190–210.

Bulkeley, H. and Stripple, J. (2013) 'Towards a critical social science of climate change?', in J. Stripple and H. Bulkeley (eds) *Governing the Climate: New Approaches to Rationality, Power and Politics*. Cambridge: Cambridge University Press, pp 243–260.

Carling, J. and Schewel, K. (2018) 'Revisiting aspiration and ability in international migration', *Journal of Ethnic and Migration Studies*, 44(6): 945–963.

Cattaneo, C., Beine, M., Fröhlich, C.J., Kniveton, D., Martinez-Zarzoso, I., Mastrorillo, M., Millock, K., Piguet, E. and Schraven, B. (2019) 'Human migration in the era of climate change', *Review of Environmental Economics and Policy*, 13(2): 189–206.

Chen, J.J., Mueller, V., Jia, Y. and Tseng, S.K.H. (2017) 'Validating migration responses to flooding using satellite and vital registration data', *American Economic Review*, 107(5): 441–445.

Christoplos, I., Ngoan, L.D., Sen, L.T.H., Huong, N.T.T. and Lindegaard, L.S. (2017) 'The evolving local social contract for managing climate and disaster risk in Vietnam', *Disasters*, 41(3): 448–467.

Clement, V., Rigaud, K.K., de Sherbinin, A., Jones, B., Adamo, S., Schewe, J., Sadiq, N. and Shabahat, E. (2021) *Groundswell Part 2: Acting on Internal Climate Migration*. Washington DC: World Bank.

Crawley, S., Coffé, H. and Chapman, R. (2022) 'Climate belief and issue salience: comparing two dimensions of public opinion on climate change in the EU', *Social Indicators Research*, 162: 307–325.

Cresswell, T. (2010) 'Towards a politics of mobility', *Environment and Planning D: Society and Space*, 28(1): 17–31.

De Haas, H. (2021) 'A Theory of migration: the aspirations-capabilities framework', *Comparative Migration Studies*, 9(1): 1–35.

De Sherbinin, A., Grace, K., McDermid, S., van der Geest, K. and Bell, A. (2022) 'Migration theory in climate mobility research', *Frontiers*, 4.

Eriksen, S.H., Nightingale, A.J. and Eakin, H. (2015) 'Reframing adaptation: the political nature of climate change adaptation', *Global Environmental Change*, 35: 523–533.

Eriksen, S.H. et al (2021) 'Adaptation interventions and their effect on vulnerability in developing countries: help, hindrance or irrelevance?', *World Development*, 141: 105383.

Ferris, E. (2020) 'Research on climate change and migration: where are we and where are we going?', *Migration Studies*, 8(4): 612–625.

Foresight (2011) *Foresight: Migration and Global Environmental Change, Final Project Report*. London: Government Office for Science.

Gebre, G.G. and Rahut, D.B. (2021) 'Prevalence of household food insecurity in East Africa: linking food access with climate vulnerability', *Climate Risk Management*.

Geddes, A., Adger, N., Arnell, N.W., Black, R. and Thomas, D. (2012) 'The implications for governance of migration linked to environmental change: key findings and new research directions', *Environment and Planning C: Government and Policy*, 30: 1078–1082.

Gemenne, F. (2015) 'One good reason to speak of 'climate refugees'', *Forced Migration Review*, 49: 70–71.

Hoffmann, R., Dimitrova, A., Muttarak, R., Crespo Cuaresma, J. and Peisker, J. (2020) 'A meta-analysis of country-level studies on environmental change and migration', *Nature Climate Change*, 10(10): 904–912.

Hub, K. (2020) 'Why focus on children: a literature review of child-centred climate change adaptation approaches', *Australian Journal of Emergency Management*, 35(2): 26–33.

Hugo, G. (2011) 'Future demographic change and its interactions with migration and climate change', *Global Environmental Change*, 21: S21–S33.

Hunter, L.M. and Simon, D.H. (2023) 'Time to mainstream the environment into migration theory?', *International Migration Review*, 57(1): 5–35.

Hunter, L.M., Luna, J.K. and Norton, R.M. (2015) 'Environmental dimensions of migration', *Annual Review of Sociology*, 41: 377–397.

Ionesco, D., Mokhnacheva, D. and Gemenne, F. (2017) *The Atlas of Environmental Migration*. New York: Routledge.

Jakobeit, C. and Methmann, C. (2012) '"ClimateRefugees" as dawning catastrophe? A critique of the dominant quest for numbers', in J. Scheffran, M. Brzoska, H.G. Brauch, P.M. Linkand and J. Schilling (eds) *Climate Change, Human Security and Violent Conflict: Challenges for Societal Stability*. New York: Springer, pp 301–314.

Kleist, N. (2014) 'Understanding diaspora organizations in European development cooperation: approaches, challenges and ways ahead', *New Diversities*, 16(2): 55–70.

Lavenex, S. and Piper, N. (2021) 'Regions and global migration governance: perspectives "from above", "from below" and "from beyond"', *Journal of Ethnic and Migration Studies*, 1–18.

Lindegaard, L.S. (2020a) 'Lessons from climate-related planned relocations: the case of Vietnam', *Climate and Development*, 12(7): 600–609.

Lindegaard, L.S. (2020b) 'A historical, scaled approach to climate change adaptation: the case of Vietnam', *Journal of Political Ecology*, 27(1): 105–124.

Lindegaard, L.S. and Sen, L.T.H. (2022) 'Everyday adaptation, interrupted agency and beyond: examining the interplay between formal and everyday climate change adaptations', *Ecology and Society*, 27(4): 42.

Lund, C. (2006) 'Twilight institutions: public authority and local politics in Africa', *Development and Change*, 37: 685–705.

Lund, C. (2016) 'Rule and rupture: state formation through the production of property and citizenship', *Development and Change*, 47(6): 1199–1228.

McCarney, R. and Kent, J. (2020) 'Forced displacement and climate change: time for global governance', *International Journal*, 75(4): 652–661.

McLeman, R.A. and Hunter, L.M. (2010) 'Migration in the context of vulnerability and adaptation to climate change: insights from analogues', *Wiley Interdisciplinary Reviews: Climate Change*, 1(3): 450–461.

Methmann, C. and Oels, A. (2015) 'From "fearing" to "empowering" climate refugees: governing climate-induced migration in the name of resilience', *Security Dialogue*, 46(1): 51–68.

Mitchell, T. (1991) 'The limits of the state: beyond statist approaches and their critics', *American Political Science Review*, 85: 77–96.

Mthethwa, S. and Wale, E. (2023) 'Household vulnerability to climate change in South Africa: a multilevel regression model', *Development Southern Africa*, 40(2): 466–481.

Nightingale, A.J. (2017) 'Power and politics in climate change adaptation efforts: struggles over authority and recognition in the context of political instability', *Geoforum*, 84: 11–20.

Nightingale, A.J. (2018) 'The socioenvironmental state: political authority, subjects, and transformative socionatural change in an uncertain world', *Environment and Planning E: Nature and Space*, 1(4): 688–711.

Nightingale, A.J., Gonda, N. and Eriksen, S.H. (2022) 'Affective adaptation = effective transformation? Shifting the politics of climate change adaptation and transformation from the status quo', *Wiley Interdisciplinary Reviews: Climate Change*, 13(1): 1–16.

Piguet, E. (2013) 'From "primitive migration" to "climate refugees": the curious fate of the natural environment in migration studies', *Annals of the Association of American Geographers*, 103(1): 148–162.

Piguet, E. (2021) 'Linking climate change, environmental degradation, and migration: An update after 10 years', *Wiley Interdisciplinary Reviews: Climate Change*, 13(1): e746.

Porst, L. and Sakdapolrak, P. (2018) 'Advancing adaptation or producing precarity? The role of rural-urban migration and translocal embeddedness in navigating household resilience in Thailand', *Geoforum*, 97: 35–45.

Porst, L. and Sakdapolrak, P. (2020) 'Gendered translocal connectedness: rural–urban migration, remittances, and social resilience in Thailand', *Population, Space and Place*, 26(4): e2314.

Ribot, J. (2014) 'Cause and response: vulnerability and climate in the Anthropocene', *Journal of Peasant Studies*, 41(5): 667–705.

Rigaud, K.K., de Sherbinin, A., Jones, B., Bergmann, J., Clement, V., Ober, K., Schewe, J., Adamo, S., McCusker, B., Heuser, S. and Midgley, A. (2018) *Groundswell: Preparing for Internal Climate Migration*. Washington DC: World Bank.

Rockenbauch, T., Sakdapolrak, P. and Sterly, H. (2019) 'Beyond the local: exploring the socio-spatial patterns of translocal network capital and its role in household resilience in Northeast Thailand', *Geoforum*, 107: 154–167.

Rohregger, B., Bender, K., Kinuthia, B.K., Schüring, E., Ikua, G. and Pouw, N. (2021) 'The politics of implementation: the role of traditional authorities in delivering social policies to poor people in Kenya', *Critical Social Policy*, 41(3): 404–425.

Rose, N. and Miller, P. (1992) 'Political power beyond the state: problematics of government', *British Journal of Sociology*, 43(2): 173–205.

Schapendonk, J., van Liempt, I., Schwarz, I. and Steel, G. (2014) 'Re-routing migration geographies: migrants, trajectories and mobility regimes', *Geoforum*, 116: 211–216.

Seter, H., Theisen, O.M. and Schilling, J. (2018) 'All about water and land? Resource-related conflicts in East and West Africa revisited', *GeoJournal*, 83(1): 169–187.

Sheller, M. (2018) *Mobility Justice: The Politics of Movement in an Age of Extremes*. New York: Verso.

Sheller, M. and Urry, J. (2006) 'The new mobilities paradigm', *Environmental Planning A*, 38(2): 207–226.

Siddiqi, A. and Blackburn, S. (2022) 'Scales of disaster: intimate social contracts on the margins of the postcolonial state', *Critique of Anthropology*, 42(3): 324–340.

Sikor, T. and Lund, C. (2009) 'Access and property: a question of power and authority', *Development and Change*, 40(1): 1–22.

Stapleton, S.O., Nadin, R., Watson, C. and Kellett, J. (2017) *Climate Change, Migration and Displacement: The Need for a Risk-Informed and Coherent Approach*. Washington, DC: UNDP & ODI.

Sørensen, N.N. (2016) 'Migrants, remittances and hometown associations in promoting development', in J. Grugeland and D. Hammett (eds) *The Palgrave Handbook of International Development*. London: Palgrave Macmillan, pp 333–345.

Taylor, M. (2014) *The Political Ecology of Climate Change Adaptation: Livelihoods, Agrarian Change and the Conflicts of Development*. Abingdon: Routledge.

Urwin, K. and Jordan, A. (2008) 'Does public policy support or undermine climate change adaptation? Exploring policy interplay across different scales of governance', *Global Environmental Change*, 18(1): 180–191.

Van Hear, N., Bakewell, O. and Long, K. (2018) 'Push-pull plus: reconsidering the drivers of migration', *Journal of Ethnic and Migration Studies*, 44(6): 927–944.

Vinke, K., Bergmann, J., Blocher, J., Upadhyay, H. and Hoffmann, R. (2020) 'Migration as Adaptation?', *Migration Studies*, 8(4): 626–634.

Warner, K. (2010) 'Global environmental change and migration: governance challenges', *Global Environmental Change*, 20(3): 402–413.

Webster, N. (2023) 'Shaping spaces: governance and climate-related mobility in Ethiopia', *Climate and Development*. https://doi.org/10.1080/17565 529.2023.2227148

Wiegel, H., Boas, I. and Warner, J. (2019) 'A mobilities perspective on migration in the context of environmental change', *WIREs Climate Change*, 10(6).

Zetter, R. (2017) 'Why they are not refugees: climate change, environmental degradation and population displacement', *Migration – Muuttoliike*, 1: 23–28.

Zickgraf, C. (2021) 'Climate change, slow onset events and human mobility: reviewing the evidence', *Environmental Sustainability* (50): 21–30.

PART I

Climate and Land Cover Changes

2

Vegetation Cover Changes in the Eastern and Upper West Regions of Ghana

*Joseph Kofi Teye, Philip Prince Kwasi Mantey,
Marie Ladekjær Gravesen, Richard Seyram,
Lily Salloum Lindegaard, Francis Xavier Jarawura,
Nauja Kleist and Shelta Gatsey*

Introduction

Natural vegetation covers large parts of the Earth's surface and constitutes its biologically richest and most complex ecosystems (Du et al, 2015). The natural vegetation is of great importance for regional hydrology, carbon storage and global climatic processes (Forkel et al, 2013; Igbawua et al, 2016). Natural vegetation is also of great economic value as a major source of food, timber and other products, and as a source of land to enhance the survival of all living beings, particularly humans (Newbold et al, 2014).

A combination of factors has resulted in increasing vegetation cover conversion rates all over the world (Braatz, 2001; Kindu et al, 2015). Scholars have identified climate change, human activity and atmospheric CO_2 fertilization effect as key influencers of vegetation cover change (Piao et al, 2006; Xin and Xu, 2007). Vegetation response to a radically changing climate has become a central concern for many scholars (Bao et al, 2014; Palmate et al, 2017; Gu et al, 2018). Consequently, researchers have identified higher annual temperatures as a leading cause of vegetation change (Karlsen et al, 2008; Wang et al, 2011; Yuan-Dong et al, 2011) and have indicated the sensitivity of the state of vegetation growth to variations in rainfall (Hu et al, 2011; Zhang et al, 2019), particularly in arid and semi-arid areas (Tong et al, 2017; Lu et al, 2019).

Besides climate change, human activities significantly impact vegetation cover conditions (Ma et al, 2019) through land use and land cover (LULC) dynamics. Land cover refers to the physical and biological cover observed on the Earth's land surface, such as vegetation or humanmade features (Anteneh et al, 2018; Mariye et al, 2022). Land use encompasses the human arrangements and activities undertaken in a specific land cover category for both social and economic purposes, such as grazing, timber extraction and conservation (Patel et al, 2019; Watson et al, 2000). Global land use changes are rapidly transforming land cover, particularly in the tropics (Houghton, 1994). These LULC changes have been identified as primary drivers of significant changes in ecosystem services both globally and in Africa (Birhanu et al, 2019; Mohamed et al, 2020).

In recent decades, African vegetation, such as grassland, woodland and forest, has been converted into agricultural lands and settlements (Lambin and Geist, 2008; Sewnet and Abebe, 2018). According to Eva et al (2006), Africa has witnessed the disappearance of 5 per cent of its woodlands and grasslands and 16 per cent of its natural forest cover between 1975 and 2000. They also estimate that Africa loses over 50,000 km² of its natural vegetation annually, most of it converted into agricultural land use or settlements. In Ghana, it is estimated that 78 per cent of forest cover disappeared between the 1900s and 1989 (Wagner and Cobbinah, 1993). Faced with these high rates of vegetation cover change, Africa has consequently suffered from various ecological problems, including habitat fragmentation, ecosystem degradation and losses in biodiversity and vegetation resources (Lucas et al, 2015; Ofori et al, 2015; Romijn et al, 2015). All these issues have direct links with both local and global environmental and climatic changes (Ochege and Okpala-Okaka, 2017).

LULC change assessment is important for studying the relationships between humans and the environment (Alam et al, 2020). Understanding the rate and level of vegetation cover change can provide a foundation for ecological and environmental protection and restoration (Gu et al, 2018; Sun et al, 2019). While it is widely recognized that having reliable data on vegetation cover change is crucial for monitoring, planning, and managing the natural environment (Fathizad et al, 2015; Khan et al, 2020), there have been limited studies conducted on land cover changes in Ghana. Given this context, this chapter utilizes geographic information systems (GIS) analysis to examine the rate and changes of green cover in four districts in the Eastern region, located in the forest zone, and the Upper West region, located in the Northern Savannah zone of Ghana. Additionally, it adopts a political ecology perspective to discuss the causes of vegetation cover changes.

Overall, the chapter finds that the study areas have experienced significant vegetation cover loss, with multiple interacting drivers including proximate activities such as logging and mining, and important underlying factors including inappropriate government policies and poverty. Efforts to address vegetation loss and improve sustainable resource management will need to address not only proximate but also underlying factors, and involve a range of actors across scales.

Theoretical perspectives on causes of vegetation cover loss

Our review of the literature shows that theories to explain forms of environmental degradation, including vegetation cover loss in Ghana, can be put into four categories: the classical, neoliberal, political economy and political ecology perspectives. The classical view (neo-Malthusian perspective) attributes environmental degradation (including vegetation loss) to mounting demographic pressures (increasing population) (see Brown, 1989; Ehrlich and Ehrlich, 1990). The classical view has been severely criticized for a single hypothesis type of explanation (Chambers, 1994). The neoliberal perspective suggests that faulty incentives or benefit-sharing systems provided by governments are the main causes of environmental degradation. Reducing government's influence on markets and liberalising trade have been suggested as effective measures for addressing environmental deterioration (Teye, 2005). The political economy perspective attributes environmental degradation to economic forces, social relations, property rights and distribution of power (Chambers, 1994). This perspective identifies poverty and external factors (such as unequal trade) as underlying causes of environmental degradation in developing countries.

The political ecology perspective emphasizes multiple causation and multiple interventions to environmental challenges. It combines the concerns of ecology and a broadly defined political economy (Blaikie and Brookfield, 1987; Yaro, 2004). This is the main theoretical perspective adopted in this chapter to explain vegetation loss in Ghana. According to Blaikie and Brookfield (1987) and Teye (2005), the application of this approach entails examination of: (a) the relationship between resource managers and the resource; (b) resource managers and interest groups in the society; and (c) the state and the global economy. Analysis should be done across scales or levels – for example, local, national and international levels. Local-level analysis focuses on such factors as cultivation techniques, population densities and sociocultural traditions. National-level analysis considers processes of national policy formulation and implementation as they can influence resource extraction. The last level of analysis focuses on the relationship between the state (Ghana) and the global economy. Things such as international trade policies may have an influence on resource extraction and degradation.

Study regions and districts

Ghana lies within the tropical climate zone and is further divided into three distinct agro-ecological zones: the Northern Savannah zone (Sudan and Guinea Savannah), the Tropical Forest zone (rainforest, the semi-deciduous forest and the transitional forest zones) and the Coastal Savannah zone. We performed vegetation change detection analysis in the Forest zone (represented by the Eastern region) and the Northern Savannah zone (represented by the Upper West region) to compare the findings of two different zones.

The Eastern region is located between latitudes 6° and 7° North and between longitudes 1°35' West and 0°30' East in Southern Ghana (see Figure 2.1). It is more urbanized and relatively economically stronger than the Upper West region. Farmers in this region produce cocoa (Ghana's most important cash crop) and food crops such as plantain, tomato, cocoyam and cassava. Two districts, namely the Yilo Krobo and Fanteakwa districts, were selected as the main study sites from the Eastern region. The Yilo Krobo Municipality covers an area of 805 sq.km and lies within the dry equatorial climatic zone of Ghana. The geographical location is approximately between latitudes 6°00'N and 6°30'N and between longitudes 00°30'W and 00°10'E. The municipal capital, Somanya, is approximately 50 km from Accra. The Fanteakwa district lies between longitudes 00°32.5'W and 00°10'W and latitudes 6°15'N and 6°40'N with a total landmass of 1,150 sq.km (GSS, 2014).

While the Eastern region generally lies within the Forest zone, its vegetation varies slightly from one place to another. The Fanteakwa district and the Northwest part of the Yilo Krobo municipality fall within the semi deciduous Forest zone, while the Accra plains dominate the southeastern lowlands, which are covered with savannah vegetation. The semi-deciduous forest is made up of large commercial trees which are usually harvested for timber. Parts of the Forest zone are under reservation. The Fanteakwa district and a few of the communities in Yilo Krobo district fall within the Worobong forest reserve. Although agriculture is generally not allowed within forest reserves, a few portions were delineated 'as admitted farms' at the time of gazetting the reserves. Production of annual crops is also sometimes allowed within some degraded reserves as part of the 'Taungya system', in which farmers are permitted to cultivate food crops on forest lands in return for planting trees on such farms for the Forestry Department (Teye, 2008).

Figure 2.1: Map of Ghana showing the regions and study districts

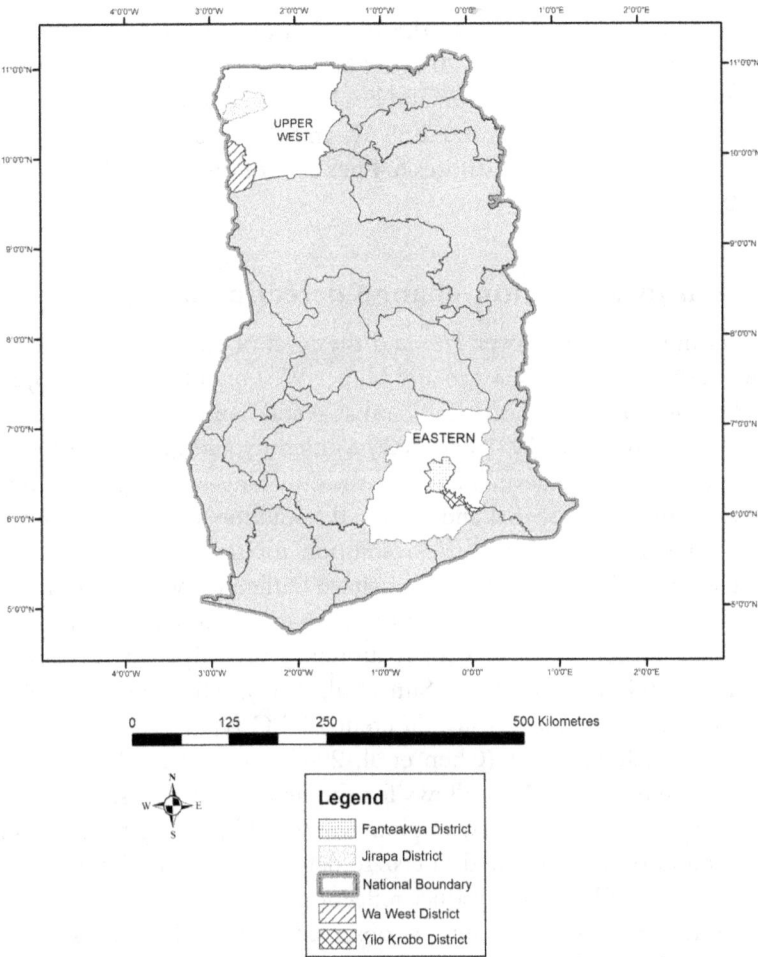

Source: Authors' construct, originally published open access in Teye et al (2021)

The Upper West region was selected to represent the semi-arid zone of Ghana. It is in the northwestern corner of Ghana and lies between longitudes 1°25'W and 2°55'W, and latitudes 9°35'N and 11°N. The region lies in the northern savannah agro-ecological zone, which is characterized by a single rainfall season (May–September). Agriculture (food crops) is the main economic activity in this region. The agricultural economy has not been strong enough to lift the population beyond the poverty line; the region records the highest poverty incidence in Ghana (GSS, 2018). The Wa West and Jirapa districts were selected as the main study sites. The Wa West district is situated in the western part of the region, between latitudes 9°40'N and 10°10'N and longitudes 2°20'W and 2°50'W. The district shares boundaries with the Sawla-Tuna-Karlba district to the south, the Wa district to the east, the Nawdoli district to the north and Ivory Coast to the west (GSS, 2012). The Jirapa municipality is in the northwestern corner of the region and lies between latitudes 10°25'N and 11°N and longitudes 2°15'W and 2°55'W, and covers an area of 1,188.6 square kilometres, which constitutes 6.4 per cent of the regional landmass (GSS, 2014).

Approach to vegetation change detection analysis

To understand the green cover rate and the level of change over time in the study districts (both in Eastern and Upper West regions), we employed GIS techniques to analyse historical Landsat-4 and Landsat-8 multispectral satellite datasets between 1991 and 2020 to identify and quantify the rate and changes of the green cover. Landsat is the only long-term digital archive with a medium to high spatial and temporal resolution as well as relatively consistent spectral and radiometric resolution for satellite imagery. The approach employed the use of the Normalized Difference Vegetation Index (NDVI), one of the most widely used indices for green cover extraction from satellite imagery as well as for spatiotemporal analysis of vegetation cover changes (Guha et al, 2018; Sun et al, 2019). The NDVI can also be used to categorize and classify different LULC types and changes by the suitable threshold values (Chen et al, 2006; Guha et al, 2018). As a standardized index, the NDVI allows for the generation of a single-band image displaying greenness (relative biomass or health) of vegetation of an area from multispectral satellite datasets. To determine the density of green cover on a patch of land, the distinct reflection in the Red (R) and near-infrared (NIR) bands by the plants are observed. Range of NDVI is −1 to +1. Higher values of NDVI refer to healthy and dense vegetation. Lower NDVI values show sparse vegetation, with values near zero and decreasing negative values indicating nonvegetated features, such as barren surfaces (rock and soil), water, snow and clouds. The study employed NDVI-derived greenness index in a postclassification change comparison. For each district, independently produced image classification results of NDVI images from the two dates of interest were obtained. This was followed by a pixel-by-pixel or segment-by-segment comparison to detect changes in the land cover classes. By adequately coding the classification results, it was possible to define the change classes and calculate transition rates between classes from the change matrix that was constructed. The main strength of this approach is that the problem of radiometric calibration between dates or using images from different satellite sensors is minimized. The ability to group classification results selectively also allows for the observation of any subset of the image. It also allows for detecting the levels and changes of greenness. Further, human errors such as pixel misclassification are avoided as its computer-based algorithm is less dependent on the analyst.

Data for change detection analysis

As noted earlier, the data for the change detection analysis were made up of available time series Level-1 Landsat-5 Thematic Mapper (TM) and Landsat-8 (OLI) images covering the study districts from 1986 to 2020, downloaded from the US Geological Survey (USGS) (https://earthexplorer.usgs.gov). Landsat Level-1 products are precision and terrain corrected, which provide for radiometric and geodetic accuracy by incorporating ground control points while employing a Digital Elevation Model (DEM) for topographic displacement. These images were viewed and the quality for potential green cover change analysis was determined. January and December had the most cloud-free and usable images, and coincided with the dry season, where subtle rainfall effects on vegetative growth are unlikely. Priority was given to anniversary and near-anniversary date images to reduce scene-to-scene variation due to sun angle, soil moisture, atmospheric conditions and vegetation phenology differences (Table 2.1). The characteristics of the different satellite sensors involved are shown in Table 2.2.

Data pre-processing

Geo-rectification of the selected images was not required as these were Level-1 Precision and Terrain (L1TP) corrected data which are within prescribed image-to-image tolerances of $\leq$ 12 m radial root mean square error (RMSE) suitable for time-series analysis. The selected images were corrected for atmospheric and radiometric distortions using radiometric rescaling coefficients provided in the product metadata file for the conversion of digital numbers (DNs) to top of atmosphere (TOA) reflectance. Images of the data sets are available in Teye et al (2021).

Table 2.1: Image datasets used for land cover change analysis

No.	Sensor	Bands	Path/Row	Date
1	Landsat 5 – TM	1–7	195/053	20 December 1986
2	Landsat 5 – TM	1–7	193/056	10 January 1991
3	Landsat 8 – OLI	1–9	195/053	31 December 2019
4	Landsat 8 – OLI	1–9	193/056	2 January 2020

Table 2.2: Satellite sensor characteristics

Sensor	Characteristics					
	Band	Spectral range (μm)	Ground resolution (m)	Temporal resolution	Inclination (degrees)	Swath width (km)
Landsat TM	1	0.45–0.53	30	16 days	98.2	185
	2	0.52–0.60	30			
	3	0.63–0.69	30			
	4	0.76–0.90	30			
	5	1.55–1.75	30			
	6	10.4–12.5	120			
	7	2.08–2.35	30			
Landsat OLI	1	0.43–0.45	30	16 days	98.2	185
	2	0.45–0.51	30			
	3	0.53–0.59	30			
	4	0.64–0.67	30			
	5	0.85–0.88	30			
	6	1.57–1.65	30			
	7	2.11–2.29	30			
	8	0.50–0.68	15			
	9	1.36–1.38	30			

NDVI computation and image classification

The NDVI was used to quantify the green cover, as it is useful in understanding vegetation density and health and assessing changes over space and time. It is calculated as a ratio between the R and NIR bands and is given as:

$$NDVI = \frac{NIR - R}{NIR + R}$$

The NDVI images for all the datasets were computed using the built-in utility within Erdas Imagine 2013 software. The NDVI images were subsequently classified using the threshold values provided by ESRI (2014), as shown in Table 2.3. Land cover classes for each independently produced image classification result were obtained. A change matrix was then constructed to define the change classes and to calculate the transition rates between classes for each pair of dates being analysed.

Table 2.3: Classification of the green cover type

NDVI range	Green cover type
< 0	Water
0.0–0.1	Built and barren areas
0.1–0.2	Low and open grassland
0.2–0.3	Shrub and grassland
0.3–0.6	Low to moderate forest
≥ 0.6	Dense forest

Interviews and focus group discussions with local communities

In addition to the GIS analysis, interviews and focus group discussions (FGDs) were conducted in selected communities in the Eastern and Upper West regions to understand farmers' experiences with land cover changes and their effects on livelihoods.

Land cover changes in the Eastern region

This section discusses land cover changes in the Eastern region, which lies in the Forest zone of Ghana.

Land cover changes in the Fanteakwa district between January 1991 and January 2020

Figure 2.2 presents the image classification maps for the Fanteakwa district for (a) January 1991 and (b) January 2020, showing the various land cover classes. Table 2.4 shows detailed statistics of the various land cover classes for the two periods under analysis. For January 1991 (Figure 2.2a), out of a total area of 1085.28 km^2, the district was covered by 157.46 km^2 (14.51 per cent) of dense forest vegetation found mainly in the middle of the western and portions of the southwestern parts. Open forest vegetation accounted for 323.56 km^2 (29.82 per cent) of the land cover found in all but the northeastern parts of the district near the Volta Lake. Thus, in all, forest vegetation accounted for about 481.02 km^2 (44.33 per cent) of the land cover in the district in 1991. Built or barren areas which represent settlements, roads and other cleared surfaces accounted for just 105.39 km^2 (9.7 per cent) of the land cover, with the rest distributed between dense shrubs/thick grassland (230.96 km^2: 21.28 per cent), low or open grassland (193.53 km^2: 17.83 per cent) and surfaces (74.36 km^2: 6.85 per cent). However, in January 2020 (Figure 2.2b), the dense forest and open forest had reduced to 90.93 km^2 (8.26 per cent) and

Figure 2.2: Land cover changes in the Fanteakwa district between (a) January 1991 and (b) January 2020

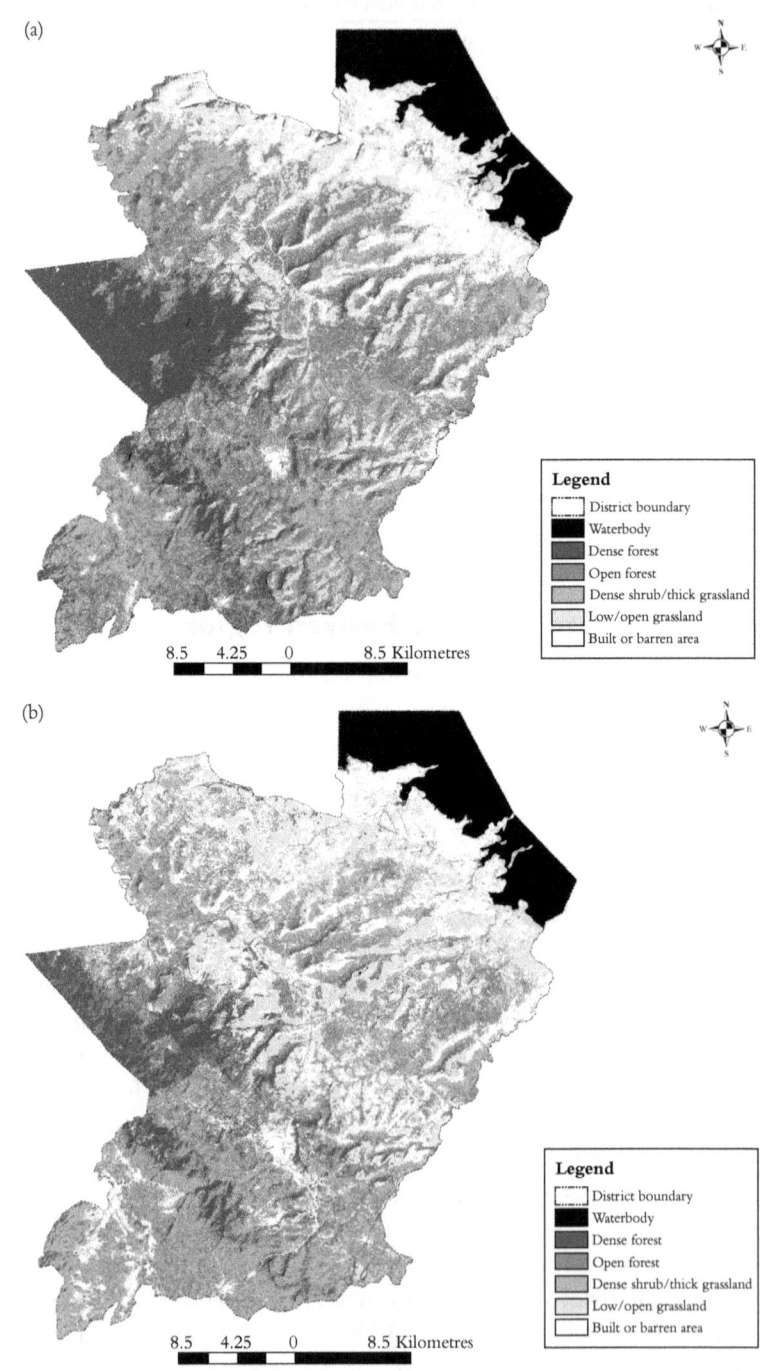

Source: Authors' construct, originally published in Teye et al (2021)

Table 2.4: Land cover changes in the Fanteakwa district between January 1991 and January 2020

10 January 1991				2 January 2020			Change (+/-)	
Land cover	Pixels	Area (km²)	%	Pixels	Area (km²)	%	Area (km²)	%
Waterbody	82,626	74.36	6.85	101,108	91.00	8.27	16.63	22.37
Dense forest	174,957	157.46	14.51	101,036	90.93	8.26	−66.53	−42.25
Open forest	359,542	323.59	29.82	219,812	197.83	17.97	−125.76	−38.86
Dense shrub/thick grassland	256,620	230.96	21.28	369,035	332.13	30.18	101.17	43.81
Low/open grassland	215,030	193.53	17.83	233,966	210.57	19.13	17.04	8.81
Built or barren area	117,097	105.39	9.71	197,968	178.17	16.19	72.78	69.06
Total	1,205,872	1,085.28	100.00	1,222,925	1,100.63	100.00		

197.83 km² (17.97 per cent) respectively, while the area covered by the dense shrub/thick grassland and low/open grassland had increased to 332.13 km² (30.18 per cent) and 210.57 km² (19.13 per cent) respectively. Also telling is the increase in built and barren areas to 178.17 km² (16.19 per cent).

Table 2.4 provides change between the various land cover classes for the Fanteakwa district between 1991 and 2020. It shows that the dense forest and open forest categories decreased by 66.53 km² (42.25 per cent) and 125.76 km² (38.86 per cent) respectively between January 1991 and January 2020. The thinning of the dense forests into open forest as well as the conversion from open forests to dense shrubs and thick grassland is clearly discernible from Figure 2.2. In all, total forest area had reduced by 192.29 km² (81.11 per cent) between the two periods. On the other hand, dense shrubs/thick grassland and low or open grassland saw an increase between the same periods by 101.17 km² (43.81 per cent) and 17.04 km² (8.81 per cent) respectively. Built and barren surfaces from increased settlement build-up, roads and other land use activities saw the largest rate of increase of 69 per cent (72.78 km²).

Land cover changes in the Yilo Krobo district between January 1991 and January 2020

Figure 2.3 presents the image classification maps for the Yilo Krobo district for (a) January 1991 and (b) January 2020, showing the various land cover classes. Table 2.5 shows the associated statistics of the various land cover classes for the two periods under analysis. For January 1991 (Figure 2.3a), out of a total area of 487.15 km², the district was covered by 138.25 km² (28.38 per cent) of dense forest vegetation found mainly in the middle to the western parts. Open forest vegetation also accounted for 229.63km² (47.14 per cent) of land cover, which was found in all but the eastern parts of the district. Thus, in all, forest vegetation accounted for about 367.88 km² (75.52 per cent) of the land cover in the district in 1991. Built or barren areas which represent settlements, roads and other cleared surfaces accounted for just 8.40 km² (1.72 per cent) of the land cover, with the rest distributed between dense shrubs/thick grassland (73.23 km²: 15.03 per cent) and low or open grassland (32.22 km²: 6.61 per cent), with water surfaces, comprising flooded areas in the eastern part, covering 5.42 km² (1.11 per cent). However, in January 2020 (Figure 2.3b), the dense forest and open forest categories had reduced to 104.76 km² (21.5 per cent) and 107.23 km² (22.01 per cent) respectively, while the area covered by the dense shrub/thick grassland and low/open grassland had increased to 195.10 km² (40.05 per cent) and 62.04 km² (12.74 per cent) respectively. Built and barren areas also saw an increase to 18.02 km² (3.7 per cent), while water surfaces had completely disappeared.

Table 2.5 also provides the rate of change between the various land cover classes for the Yilo Krobo district between 1991 and 2020. It shows that

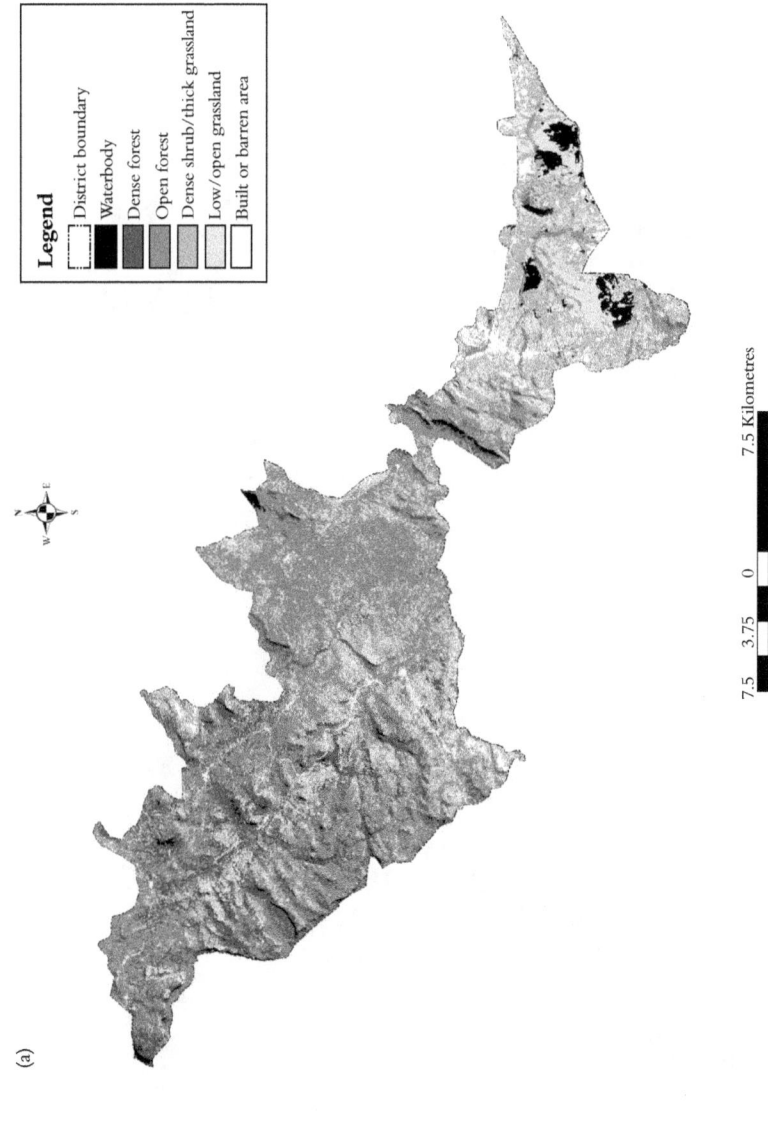

Figure 2.3: Land cover changes in the Yilo Krobo district between (a) January 1991 and (b) January 2020

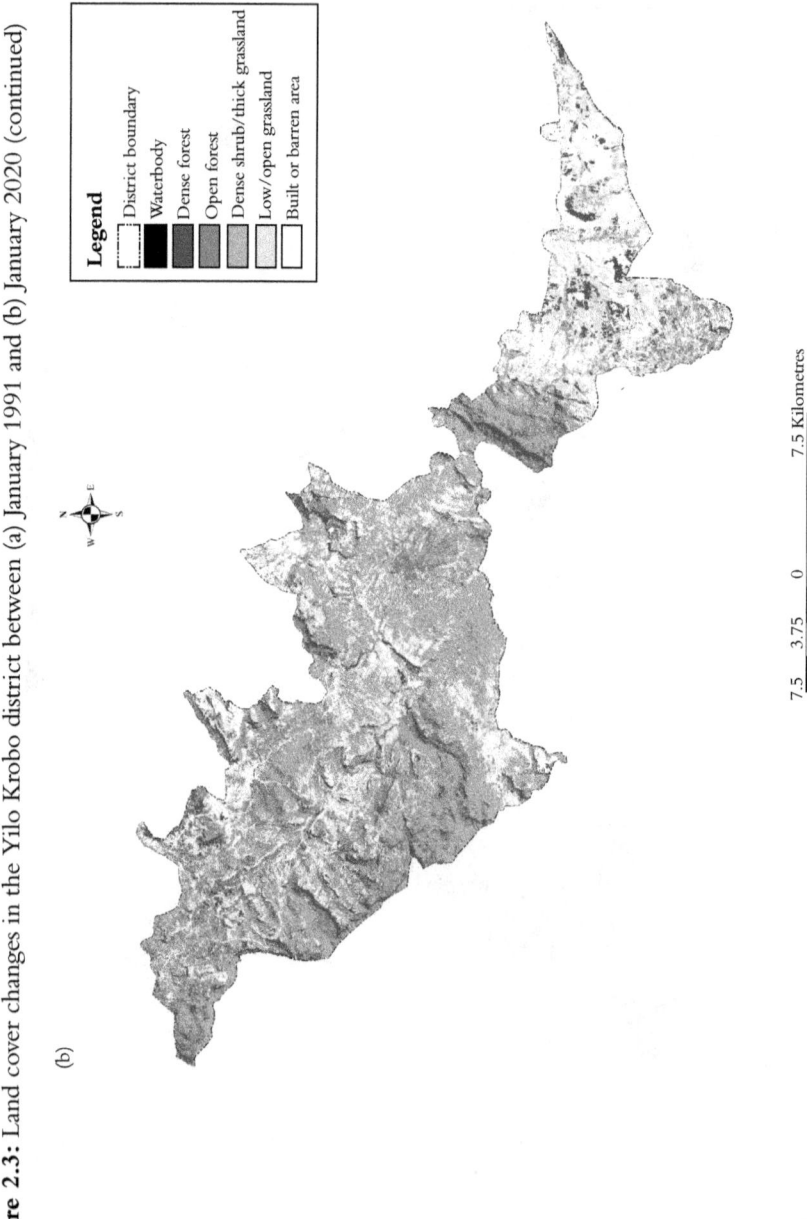

Figure 2.3: Land cover changes in the Yilo Krobo district between (a) January 1991 and (b) January 2020 (continued)

Source: Authors' construct, originally published in Teye et al (2021)

Table 2.5: Land cover changes in the Yilo Krobo district between January 1991 and January 2020

Land Cover	10 January 1992			2 January 2020			Change (+/-)	
	Pixels	Area (km²)	%	Pixels	Area (km²)	%	Area (km²)	%
Waterbody/flooded area	6,027	5.42	1.11	0	0	0	-5.42	100.00
Dense forest	153,616	138.25	28.38	116,397	104.76	21.50	-33.50	-24.23
Open forest	255,144	229.63	47.14	119,148	107.23	22.01	-122.40	-53.30
Dense shrub/thick grassland	81,362	73.23	15.03	216,775	195.10	40.05	121.87	166.43
Low/open grassland	35,797	32.22	6.61	68,938	62.04	12.74	29.83	92.58
Built or barren area	9,337	8.40	1.72	20,025	18.02	3.70	9.62	114.47
Total	541,283	487.15	100.00	541,283	487.15	100.00		

the dense forest and open forest categories decreased by 33.5 km² (24.23 per cent) and 122.40 km² (53.30 per cent) respectively between January 1991 and January 2020, and this is particularly visible in the middle of the district. In all, total forest area had reduced by 155.4 km² (77.53 per cent) between the two periods. On the other hand, dense shrubs/thick grassland and low or open grassland saw an increase between the same periods of 121.87 km² (166.43 per cent) and 29.83 km² (92.58 per cent). Built and barren surfaces from increased settlement build-up, roads and other land use activities also showed a large rate of increase of 114.47 per cent (9.62 km²), while water surfaces showed a 100 per cent decrease between January 1991 and January 2020.

Change detection analysis for the Upper West region of Ghana

The change detection was also performed for the Upper West region, using similar data and following the same procedures described in the methodology.

Land cover changes in the Jirapa district between December 1986 and December 2019

Figure 2.4 presents the classified maps for the Jirapa district for (a) December 1986 and (b) December 2019, showing the various land cover classes. Table 2.6 shows detailed statistics of the various land cover classes for the two periods under analysis. For December 1986 (Figure 2.4a), out of a total area of 1,190.13 km², the district was covered by 317.27 km² (26.66 per cent) of wooded savannah vegetation found mainly in the eastern half. Shrub and tree savannah also accounted for 258.78km² (21.74 per cent) of the land cover. Thus, in all, tree vegetation accounted for about 576.05 km² (48.4 per cent) of the land cover in the district in 1986. Built or barren areas which represent settlements, roads and other cleared surfaces accounted for just 144.73 km² (12.16 per cent) of the land cover, with the rest distributed between herbaceous savannah, which represented the largest land cover class at 354.85 km² (29.82 per cent), and wetlands covering 114.5 km² (9.62 per cent). However, in December 2019 (Figure 2.4b), the wooded savannah and shrub and tree savannah had reduced to 219.61 km² (18.45 per cent) and 131.94 km² (11.09 per cent) respectively, while the area covered by the herbaceous savannah had decreased marginally to 350.42 km² (29.44 per cent). Built and barren surfaces saw the largest increase to 420.09 km² (35.3 per cent), with wetlands showing a decrease to 68.06 km² (5.72 per cent).

Table 2.6 also provides the change between the various land cover classes for the Jirapa district between 1986 and 2019. It shows that the wooded savannah and shrub and tree savannah categories decreased by 97.66 km²

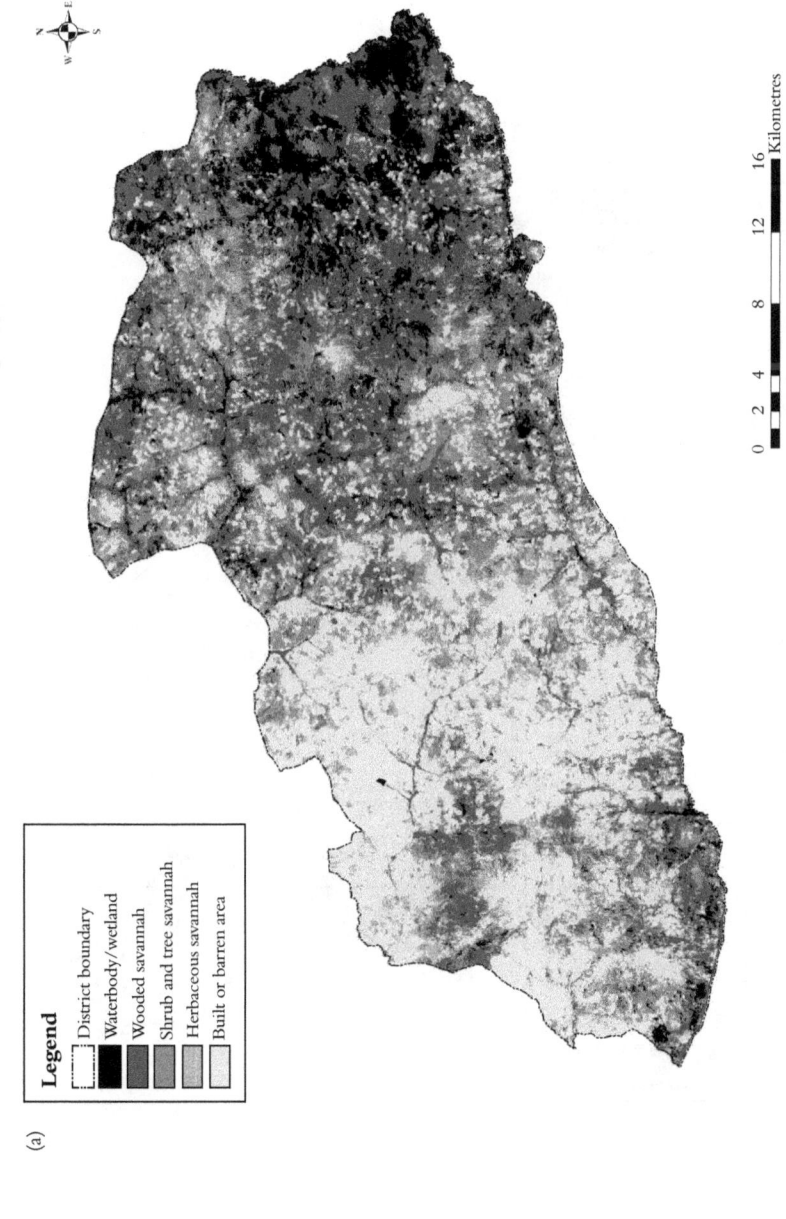

Figure 2.4: Land cover changes in the Jirapa district between (a) December 1986 and (b) December 2019

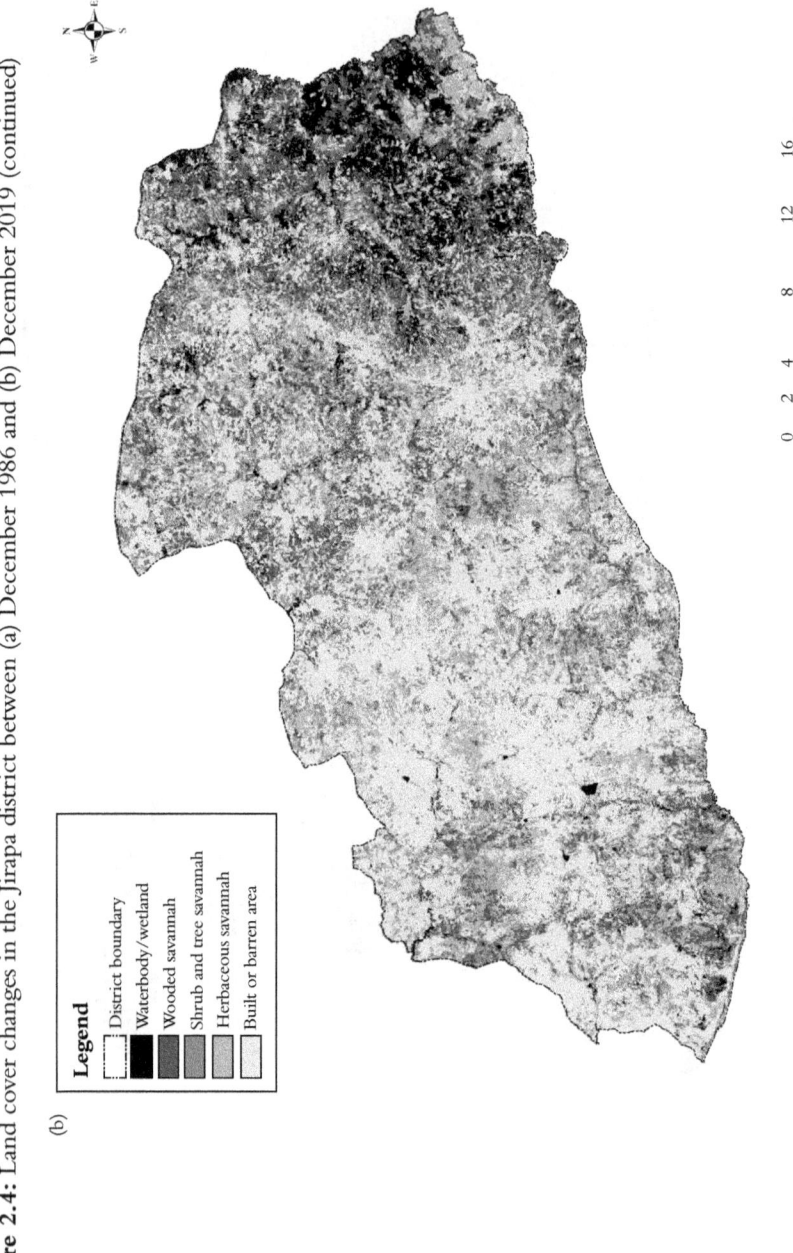

Figure 2.4: Land cover changes in the Jirapa district between (a) December 1986 and (b) December 2019 (continued)

Source: Authors' construct, originally published in Teye et al (2021)

Table 2.6: Land cover changes in the Jirapa district between December 1986 and December 2019

Land cover	20 December 1986			31 December 2019			Change (+/-)	
	Pixels	Area (km²)	%	Pixels	Area (km²)	%	Area (km²)	%
Water body/wetland	127,217	114.50	9.62	75,617	68.06	5.72	-46.44	-40.56
Wooded savannah	352,522	317.27	26.66	244,010	219.61	18.45	-97.66	-30.78
Shrub and tree savannah	287,535	258.78	21.74	146,605	131.94	11.09	-126.84	-49.01
Herbaceous savannah	394,274	354.85	29.82	389,361	350.42	29.44	-4.42	-1.25
Built or barren area	160,815	144.73	12.16	466,770	420.09	35.30	275.36	190.25
Total	1,322,363	1,190.13	100.00	1,322,363	1,190.13	100.00		

(30.78 per cent) and 126.84 km² (49.01 per cent) respectively between December 1986 and December 2019. In all, wooded and tree vegetation decreased by 224.5 km² between the two periods. Herbaceous savannah saw a marginal decrease in the same period by 4.42 km² (1.25 per cent) and 17.04 km², whereas built and barren areas saw the largest increase of 275.36 km² (190.25 per cent).

Land cover changes in the Wa West district between December 1986 and December 2019

Figure 2.5 presents the classified maps for the Wa West district for (a) December 1986 and (b) December 2019, showing the various land cover classes. Table 2.7 shows detailed statistics of the various land cover classes for the two periods under analysis. For December 1986 (Figure 2.5a), out of a total area of 1496.76 km², the district was covered by 532.83 km² (35.67 per cent) of wooded savannah vegetation in all parts but the northern end. Shrub and tree savannah also accounted for 527.23 km² (35.30 per cent) of the land cover. Thus, in all, tree vegetation accounted for about 1,060.06 km² (70.97 per cent) of the land cover in the district in 1986. Built or barren areas which represent settlements, roads and other cleared surfaces accounted for just 99.56 km² (6.67 per cent) of the land cover, with the rest distributed between herbaceous savannah at 261.80 km² (17.53 per cent) and wetlands covering 72.34 km² (4.84 per cent). However, in December 2019 (Figure 2.5b), the wooded savannah and shrub and tree savannah had reduced to 389.79 km² (26.09 per cent) and 382.21 km² (25.59 per cent) respectively, while the area covered by herbaceous savannah had increased to 302.74 km² (20.27 per cent). Built and barren surfaces saw the largest increase to 368.22 km² (24.65 per cent), with wetlands showing a decrease to 50.80 km² (3.40 per cent).

Table 2.7 also provides the change between the various land cover classes for the Wa West district between 1986 and 2019. It shows that the wooded savannah and shrub and tree savannah categories decreased by 143.04 km² (26.85 per cent) and 145.02 km² (27.51 per cent) respectively between December 1986 and December 2019. In all, wooded and tree vegetation had reduced by 288.06 km² (57.36 per cent) in the period of analysis. Herbaceous savannah also saw a decrease in the same period by 40.94 km² (15.64 per cent), whereas built and barren surfaces saw the largest rate of increase of 268.22 km² (269.84 per cent). Wetlands showed a decrease of 21.54km² or 29.77 per cent.

Discussing the causes and effects of vegetation cover loss

In the Forest zone, both key informants and farmers reported significant deforestation in recent years. At Ahinkwa, a farming community in the Yilo

Figure 2.5: Land cover changes in the Wa West district between (a) December 1986 and (b) December 2019

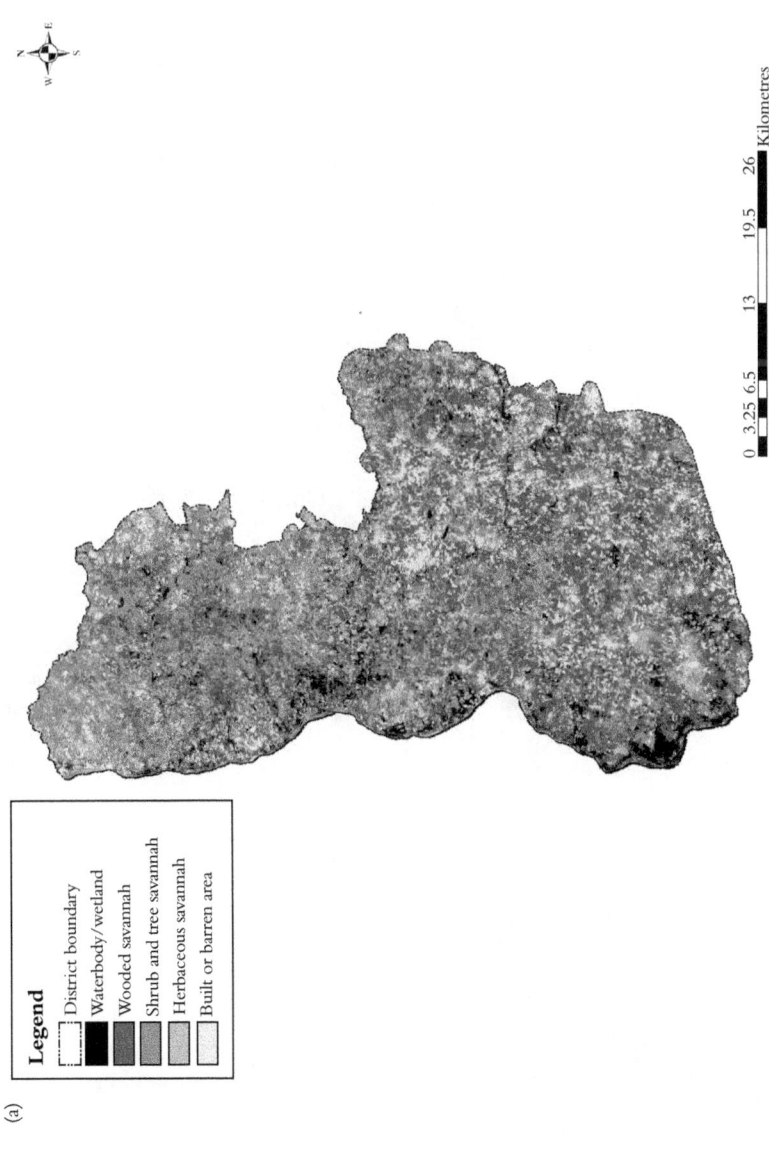

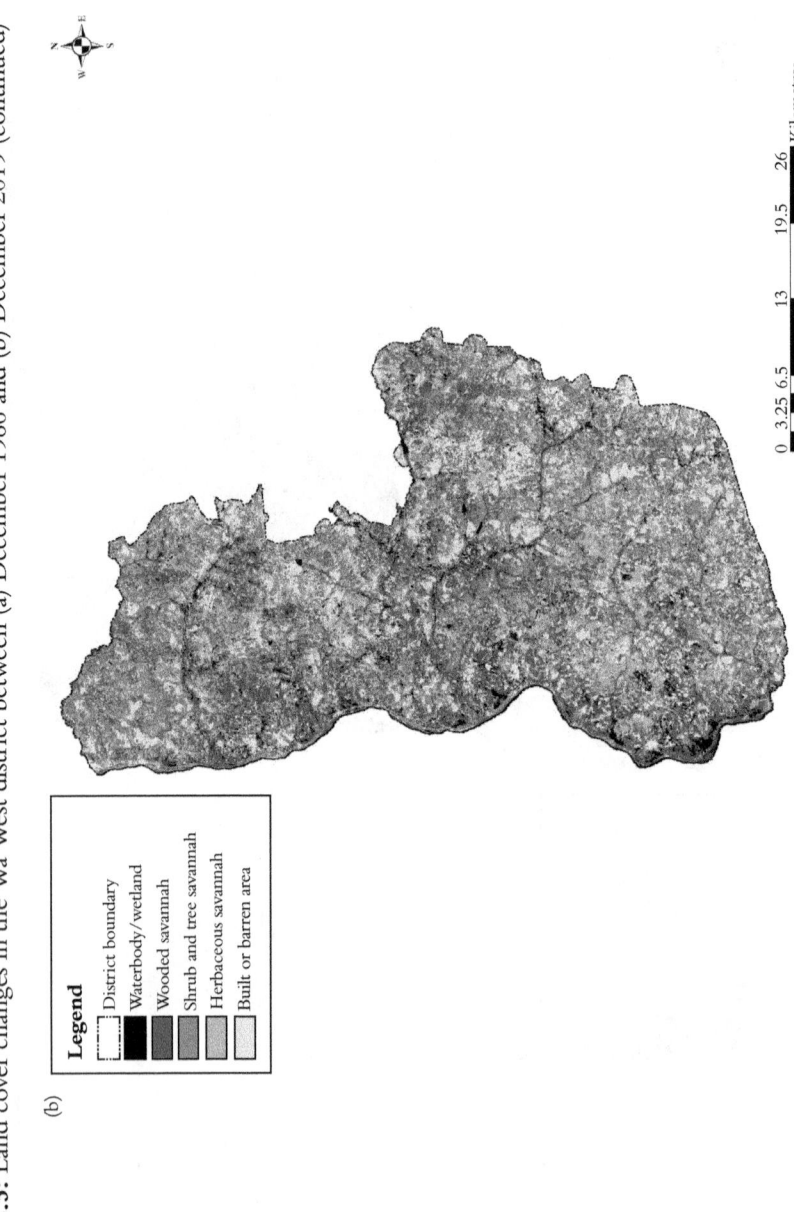

Figure 2.5: Land cover changes in the Wa West district between (a) December 1986 and (b) December 2019 (continued)

Source: Authors' construct, originally published in Teye et al (2021)

Table 2.7: Land cover changes in the Wa West district between December 1986 and December 2019

Land cover	20 December 1986			31 December 2019			Change (+/-)	
	Pixels	Area (km²)	%	Pixels	Area (km²)	%	Area (km²)	%
Water body/wetland	80,381	72.34	4.84	56,448	50.80	3.40	21.54	-29.77
Wooded savannah	592,029	532.83	35.67	433,095	389.79	26.09	143.04	-26.85
Shrub and tree savannah	585,813	527.23	35.30	424,680	382.21	25.59	145.02	-27.51
Herbaceous savannah	290,885	261.80	17.53	336,375	302.74	20.27	-40.94	15.64
Built or barren area	110,625	99.56	6.67	409,135	368.22	24.65	-268.66	269.84
Total	1,659,733	1,493.76	100.00	1,659,733	1,493.76	100.00		

Krobo district, for instance, farmers narrated how the forest cover has been depleted because of human activities. One described it as follows:

> Here was a thick forest which was protected by the Forestry Department. The situation changed in the 1980s when we [farmers] were allowed to enter the forest to farm on the forest land and plant trees. In less than ten years, we destroyed the entire forest through farming and harvesting of trees. (Farmer member of FGD)

This statement on rapid deforestation in the 1990s is consistent with the high level of deforestation in Ghana at that time. According to the Food and Agriculture Organization of the United Nations (FAO) (2010), Ghana lost an average of 125,400 hectares of forest cover, or 1.7 per cent per year, and a total of 33.7 per cent for the period (2,508,000 hectares). The deforestation affected both reserve and off-reserve forests (Teye, 2012). An assessment of forest cover by Hawthorne and Abu Juam carried out in 1993 shows that due to human activities, only 2 per cent of the total area of forest reserves was in a 'very excellent' condition (Hawthorne and Abu Juam, 1995). The assessment further shows that of the 214 forest reserves in the country, 121 (representing about half of the entire reserves area) were seriously degraded or without forests at all. Like the observation in the Eastern region, farmers in the Upper West region also explained that they have witnessed rapid vegetation cover loss in recent years. According to Teye (2005), vegetation cover loss is caused by a combination of proximate and underlying factors. Proximate factors include logging, mining activities, agricultural activities, fuel wood and charcoal production, and clearing of vegetation for human settlements. In the political ecologist perspective, while these proximate factors occur largely in the communities where vegetation loss takes place, they are triggered by far-reaching underlying factors such as rapid population increases, poverty and inappropriate policies.

Logging has been part of the Ghanaian economy since colonial times, but has worsened since the 1970s due to the activities of illegal chainsaw operators and large logging firms. As postulated by the political ecologists, faulty government policies are often blamed for the overlogging in Ghana. As a result of patronage networks between state politicians and big timber contractors, the latter pressured policy makers in 1970 to reduce the felling cycle from 25 years to 15 years, asserting that this would help prevent 'over mature' trees from 'going to waste'. This continued until 1990 before it was changed to 40 years (Prah, 1994). Again, only a small percentage of royalties (38 per cent) is paid to the state and, as such, timber companies have incentives to harvest wood indiscriminately. In 1993, for instance, total forest revenue was estimated at $5 per cubic metre of wood, although the actual value of wood at that time was $98 per cubic metre (Mayers and

Kotey, 1996). The World Bank in 1986 noted that forest resources in Ghana could be considered a 'free good' because of the exceptionally low royalty and fee levels (IBRD, 1986). Also, fines for illegal logging have historically been so low that they do not deter people from harvesting wood illegally. For instance, fines imposed for illegal harvesting in 1974 were not revised until 1994, even though inflation increased more than 1,000 per cent during this period (see ISSER, 1995; Teye, 2008). Additionally, lucrative trees on both forest reserves and off-reserve lands are legally said to be the property of the state. The royalties are paid to persons who hold land titles (usually chiefs). Consequently, local farmers who nurture trees do not benefit much from timber harvested by large timber firms. As a result, local farmers tend to support illegal chainsaw operators to harvest wood illegally (Teye, 2011). Logging also paves the way for further degradation by farmers.

Mining of gold, bauxite, diamond and manganese also contributes to vegetation loss in Ghana. Only a small proportion of royalties go to the state and, as such, mining companies continue to extract large volumes of minerals. At the same time, local people who do not benefit much from the distribution of royalties team up with artisanal local and Chinese miners, destroying large tracts of lands and polluting river bodies as part of mining activities. Consistent with the political ecologist perspective, the high demand for minerals on the international market explains this high level of illegal mining. Indeed, since 2000, thousands of Chinese miners have migrated irregularly to Ghana to engage in artisanal mining which is reserved for Ghanaians (Botwe et al, 2019; Teye et al, 2022). Additionally, with about 60 per cent of Ghanaians engaged in agriculture, significant portions of forest lands are cleared annually to produce crops. Interviews indicated that inability of farmers to buy fertilizers means they still practise a kind of shifting cultivation or bush fallowing which destroys the lands, as highlighted by an agriculture extension officer, who said that 'the vegetation is constantly destroyed by farmers because in this country, many farmers do not use fertilizer. They therefore move to clear new lands once the fertility of soils on their existing farms declines'.

Cutting of wood for firewood and charcoal is also a major cause of vegetation loss. Fuel wood is the major source of energy for rural households that constitute about 60 per cent of the population. Even in urban areas, about 69 per cent of all households use charcoal. In both the Eastern and Upper West regions, charcoal production has become a major economic activity for poor households. Additionally, as Ghana's population increases, large tracts of vegetation cover are destroyed annually for construction of settlements and roads.

It is important to emphasize that local people who were interviewed recognized the interaction between the proximate factors and underlying factors, as highlighted by the following respondent in the Eastern region:

> This place used to be forested but in the 1990s we lost the forest because of farming activities. The population increased and more farms were created which destroyed the land. Also, many people migrated to this area to farm because the land was fertile. Some did not have lands and so the forest was destroyed.

This statement draws attention to how increased population brought about by both natural increase and migration contributed to forest loss. Another respondent spoke about how poverty, which is a national problem, made it difficult to control human activities linked to deforestation:

> Now there is no forest in this community and all the valuable trees have been cut down for timber. People have now turned to charcoal production which was not an [economic] activity here about 30 years ago ... [but] poverty has now forced several people to engage in charcoal production which involves the cutting down of small trees. We had several meetings in this community to discuss how we can reduce the cutting down of young trees, but our bye laws do not work because many people are poor and there is no alternative source of livelihoods for some.

In addition to identifying the causes of vegetation cover changes, respondents also emphasize both ecological and socioeconomic impacts of the loss of vegetation cover. Consistent with the existing literature (see Brown, 1992; FAO, 1993; Teye, 2005), the ecological effects of vegetation cover loss identified by community members include climatic changes, soil erosion, depletion of water bodies and biodiversity loss. Furthermore, community members indicated that the loss of vegetation cover adversely affects their livelihoods in multiple aspects. First, declining crop production, due to fluctuating rainfall and deforestation, and vegetation cover loss was also seen to cause the depletion of water bodies, resulting in water scarcity in communities that rely on rivers for water. Vegetation change is also seen to be resulting in the loss of animals and tree species, which is negatively affecting ecotourism; the aesthetic appeal of tourist sites are being reduced due to reduced biodiversity. Hardwood, which is used for furniture and construction, is becoming scarce in Ghana, affecting the housing industry. Vegetation loss is also changing insect habitats and creating conditions for the spread of diseases. For instance, river blindness is transmitted by cytoforms of a blackfly found in savannah regions. These cytoforms are spreading into areas that were once forested but are now grassland (Teye, 2005). Vegetation loss is also linked to pharmacopoeia as many Ghanaians rely on herbal medicines. Numerous indigenous trees and plants have not been scientifically catalogued, but there are concerns that deforestation will result in the loss of these resources before they can be fully catalogued.

Concluding remarks

This chapter clearly demonstrates that both the Eastern and Upper West regions have experienced significant vegetation cover loss over the study period due to a combination of underlying and proximate factors. Proximate factors include logging, mining, agricultural activities, wood cutting for firewood and charcoal burning, and land clearance for construction purposes. Underlying factors such as demographic pressures, inappropriate government policies and poverty further exacerbate the problem. From a political ecologist perspective, it is crucial to account for this multiple causation when implementing measures to address this issue. Collaboration with development partners is necessary in the development of programs to reduce poverty, which in turn alleviates pressure on land and vegetation. Additionally, it is imperative for the Ghanaian government to review the existing resource benefit-sharing formula to ensure that communities also benefit from trees and natural resources on their lands. This will encourage sustainable resource management in practice.

References

Alam, A., Bhat, M.S. and Maheen, M. (2020) 'Using Landsat satellite data for assessing the land use and land cover change in Kashmir valley', *GeoJournal*, 85(6): 1529–1543.

Anteneh, Y., Stellmacher, T., Zeleke, G., Mekuria, W. and Gebremariam, E. (2018) 'Dynamics of land change: insights from a three-level intensity analysis of the Legedadie-Dire catchments, Ethiopia', *Environmental Monitoring and Assessment*, 190(5): 1–22.

Bao, G., Qin, Z., Bao, Y., Zhou, Y., Li, W. and Sanjjav, A. (2014) 'NDVI-based long-term vegetation dynamics and its response to climatic change in the Mongolian Plateau', *Remote Sensing*, 6(9): 8337–8358.

Birhanu, L., Hailu, B.T., Bekele, T. and Demissew, S. (2019) 'Land use/land cover change along elevation and slope gradient in highlands of Ethiopia', *Remote Sensing Applications: Society and Environment*, 16: 100260.

Blaikie, P. and Brookfield, H. (1987) *Land Degradation and Society*. London: Routledge.

Botwe, B.P., Amoah-Binfoh, K. and Masih, E. (2019) 'Organizational climate and its influence on work engagement at Sam Higginbottom University of Agriculture, Technology and Sciences (SHUATS), Allahabad, India', *All Nations University Journal of Applied Thought (ANUJAT)*, 7(1): 42–54.

Braatz, S.M. (2001) *The State of the World's Forests 2001*. Rome: Food and Agriculture Organization of the United Nations.

Brown, A.L. (1989) 'Analogical learning and transfer: what develops?', in S. Vosniadou and A. Ortony (eds) *Similarity and Analogical Reasoning*. Cambridge: Cambridge University Press, pp 369–412.

Brown, A.L. (1992) 'Design experiments: theoretical and methodological challenges in creating complex interventions in classroom settings', *Journal of the Learning Sciences*, 2(2): 141–178.

Chambers, R. (1994) 'Participatory rural appraisal (PRA): challenges, potentials and paradigm', *World Development*, 22(10): 1437–1454.

Chen, X.-L., Zhao, H.-M., Li, P.-X. and Yin, Z.-Y. (2006) 'Remote sensing image-based analysis of the relationship between urban heat island and land use/cover changes', *Remote Sensing of Environment*, 104(2): 133–146.

Du, J. et al (2015) 'Analysis on spatio-temporal trends and drivers in vegetation growth during recent decades in Xinjiang, China', *International Journal of Applied Earth Observation and Geoinformation*, 38: 216–228.

Ehrlich, P.R. and Ehrlich, A.H. (1990) *The Population Explosion*. New York: Simon & Schuster.

ESRI (Environmental Systems Research Institute) (2014). *ArcGIS 10.3 Help: NDVI Function*. Redlands: ESRI Press.

Eva, H.D., Brink, A. and Simonetti, D. (2006) 'Monitoring land cover dynamics in Sub-Saharan Africa'. Luxembourg: Institute for Environmental and Sustainability, European Communities, pp 1–44.

FAO (Food and Agriculture Organization) (1993) *Forest Resources Assessment 1990: Tropical Countries*. FAO Forestry Paper No. 112. Rome: Food and Agriculture Organization of the United Nations.

FAO (2010) *Global Forest Resources Assessment 2010: Main Report*. Rome: Food and Agriculture Organization of the United Nations.

Fathizad, H., Rostami, N. and Faramarzi, M. (2015) 'Detection and prediction of land cover changes using Markov chain model in semi-arid rangeland in western Iran', *Environmental Monitoring and Assessment*, 187(10): 1–12.

Forkel, M., Carvalhais, N., Verbesselt, J., Mahecha, M.D., Neigh, C.S. and Reichstein, M. (2013) 'Trend change detection in NDVI time series: effects of inter-annual variability and methodology', *Remote Sensing*, 5(5): 2113–2144.

GSS (Ghana Statistical Service) (2012) *2010 Population and Housing Census*. Accra: Ghana Statistical Service.

GSS (2014) *2010 Population and Housing Census*. Accra: Ghana Statistical Service.

GSS (2018) *Ghana Living Standards Survey Round 7 (GLSS7): Main Report*. Accra: Ghana Statistical Service.

Gu, Z., Duan, X., Shi, Y., Li, Y. and Pan, X. (2018) 'Spatiotemporal variation in vegetation coverage and its response to climatic factors in the Red River Basin, China', *Ecological Indicators*, 93: 54–64.

Guha, S., Govil, H., Dey, A. and Gill, N. (2018) 'Analytical study of land surface temperature with NDVI and NDBI using Landsat 8 OLI and TIRS data in Florence and Naples city, Italy', *European Journal of Remote Sensing*, 51(1): 667–678.

Hawthorne, W. and Abu-Juam, M. (1995) *Forest Protection in Ghana: With Particular Reference to Vegetation and Plant Species*. IUCN. Gland, Switzerland and Cambridge, UK.

Houghton, R.A. (1994) 'The worldwide extent of land-use change', *BioScience*, 44(5): 305–313.

Hu, M.Q., Mao, F., Sun, H. and Hou, Y.Y. (2011) 'Study of normalized difference vegetation index variation and its correlation with climate factors in the three-river-source region', *International Journal of Applied Earth Observation and Geoinformation*, 13(1): 24–33.

Igbawua, T., Zhang, J., Chang, Q. and Yao, F. (2016) 'Vegetation dynamics in relation with climate over Nigeria from 1982 to 2011', *Environmental Earth Sciences*, 75(6): 1–16.

Institute of Statistical, Social and Economic Research (1995) *The State of the Ghanaian Economy in 1994*. Accra: ISSER, University of Ghana.

International Bank of Reconstruction and Development (1986) *The Study of the Substitution of Labour and Equipment in Civil Construction: A Research and Implementation Project Completion Report Washington*. Washington DC: World Bank Group.

Karlsen, S.R., Tolvanen, A., Kubin, E., Poikolainen, J., Johansen, B., Danks, F.S. and Makarova, O. (2008) 'MODIS-NDVI-based mapping of the length of the growing season in northern Fennoscandia', *International Journal of Applied Earth Observation and Geoinformation*, 10(3): 253–266.

Khan, Z., Saeed, A. and Bazai, M.H. (2020) 'Land use/land cover change detection and prediction using the CA-Markov model: a case study of Quetta city, Pakistan', *Journal of Geography and Social Sciences*, 2(2): 164–182.

Kindu, M., Schneider, T., Teketay, D. and Knoke, T. (2015) 'Drivers of land use/land cover changes in Munessa-Shashemene landscape of the south-central highlands of Ethiopia', *Environmental Monitoring and Assessment*, 187(7): 1–17.

Lambin, E.F. and Geist, H.J. (eds) (2008) *Land-Use and Land-Cover Change: Local Processes and Global Impacts*. Springer Science & Business Media. Germany.

Lu, Q., Zhao, D., Wu, S., Dai, E. and Gao, J. (2019) 'Using the NDVI to analyze trends and stability of grassland vegetation cover in Inner Mongolia', *Theoretical and Applied Climatology*, 135(3): 1629–1640.

Lucas, R. et al (2015) 'The earth observation data for habitat monitoring (EODHaM) system', *International Journal of Applied Earth Observation and Geoinformation*, 37: 17–28.

Ma, Q., Long, Y., Jia, X., Wang, H. and Li, Y. (2019) 'Vegetation response to climatic variation and human activities on the Ordos Plateau from 2000 to 2016', *Environmental Earth Sciences*, 78(24): 1–15.

Mariye, M., Maryo, M. and Li, J. (2022) 'The study of land use and land cover (LULC) dynamics and the perception of local people in Aykoleba, northern Ethiopia', *Journal of the Indian Society of Remote Sensing*, 50(5): 775–789.

Mayers, J. and Kotey, N.A. (1996) *Local Institutions and Adaptive Management in Ghana. IIED Forestry and Land Use Series 7*. London: IIED.

Mohamed, M.A., Anders, J. and Schneider, C. (2020) 'Monitoring of changes in land use/land cover in Syria from 2010 to 2018 using multitemporal Landsat imagery and GIS', *Land*, 9(7): 226.

Newbold, T., Hudson, L.N., Phillips, H.R., Hill, S.L., Contu, S., Lysenko, I. and Purvis, A. (2014) 'A global model of the response of tropical and sub-tropical forest biodiversity to anthropogenic pressures', *Proceedings of the Royal Society B: Biological Sciences*, 281(1792): 20141371.

Ochege, F.U. and Okpala-Okaka, C. (2017) 'Remote sensing of vegetation cover changes in the humid tropical rainforests of Southeastern Nigeria (1984–2014)', *Cogent Geoscience*, 3(1): 1307566.

Ofori, B.Y., Owusu, E.H. and Attuquayefio, D.K. (2015) 'Ecological status of the Mount Afadjato–Agumatsa range in Ghana after a decade of local community management', *African Journal of Ecology*, 53(1): 116–120.

Palmate, S.S., Pandey, A., Kumar, D., Pandey, R.P. and Mishra, S.K. (2017) 'Climate change impact on forest cover and vegetation in Betwa Basin, India', *Applied Water Science*, 7(1): 103–114.

Patel, S.K., Verma, P. and Shankar Singh, G. (2019) 'Agricultural growth and land use land cover change in peri-urban India', *Environmental Monitoring and Assessment*, 191(9): 1–17.

Piao, S., Mohammat, A., Fang, J., Cai, Q. and Feng, J. (2006) 'NDVI-based increase in growth of temperate grasslands and its responses to climate changes in China', *Global Environmental Change*, 16(4): 340–348.

Prah, E.A. (1994) *Sustainable Management of the Tropical High Forest of Ghana*. London: Commonwealth Secretariat.

Romijn, E., Lantican, C.B., Herold, M., Lindquist, E., Ochieng, R., Wijaya, A. and Verchot, L. (2015) 'Assessing change in national forest monitoring capacities of 99 tropical countries', *Forest Ecology and Management*, 352: 109–123.

Sewnet, A. and Abebe, G. (2018) 'Land use and land cover change and implication to watershed degradation by using GIS and remote sensing in the Koga watershed, Northwestern Ethiopia', *Earth Science Informatics*, 11: 99–108.

Sun, R., Chen, S., Su, H. and Hao, G. (2019) 'Spatiotemporal variation of vegetation coverage and its response to climate change before and after implementation of grain for green project in the Loess Plateau', *IEEE International Geoscience and Remote Sensing Symposium*: 9546–9549.

Teye, J.K. (2005) 'Deforestation in Ghana', *Human Landscape Ecology*: 9–21.

Teye, J.K. (2008) 'Forest resource management in Ghana: an analysis of policy and institutions', Doctoral dissertation, University of Leeds.

Teye, J.K. (2011) 'Ambiguities of forest management decentralization in Ghana', *Journal of Natural Resources Policy Research*, 3(4): 355–369.

Teye, J.K. (2012) 'Benefits, challenges, and dynamism of positionalities associated with mixed methods research in developing countries: evidence from Ghana', *Journal of Mixed Methods Research*, 6(4): 379–391.

Teye, J.K., Jarawura, F., Lindegaard, L.S., Kleist, N., Ladekjær Gravesen, M., Mantey, P. and Quaye, D. (2021) *Climate Mobility: Scoping Study of Two Localities in Ghana*. DIIS Working Paper 2021: 10. Copenhagen: Danish Institute for International Studies.

Teye, J.K., Kandilige, L., Setrana, M. and Yaro, J. (2022) 'Chinese migration to Ghana: challenging the orthodoxy on characterizing migrants and reasons for migration', *Ghana Journal of Geography*, 14(2): 203–234.

Tong, S., Zhang, J. and Bao, Y. (2017) 'Spatial and temporal variations of vegetation cover and the relationships with climate factors in Inner Mongolia based on GIMMS NDVI3g data', *Journal of Arid Land*, 9(3): 394–407.

Wagner, M.R. and Cobbinah, J.R. (1993) 'Deforestation and sustainability in Ghana', *Journal of Forestry*, 91(6): 35–39.

Wang, X., Piao, S., Ciais, P., Li, J., Friedlingstein, P., Koven, C. and Chen, A. (2011) 'Spring temperature change and its implication in the change of vegetation growth in North America from 1982 to 2006', *Proceedings of the National Academy of Sciences*, 108(4): 1240–1245.

Watson, R.T., Noble, I.R., Bolin, B., Ravindranath, N.H., Verardo, D.J. and Dokken, D.J. (2000) *Land Use, land-Use Change and Forestry*. IPPC Report.

Xin, Z.B. and Xu, J.X. (2007) 'Spatial and temporal evolution of vegetation cover in the Loess Plateau and its response to climate change', *Progress in Natural Science*, 17(6): 770–778.

Yaro, J.A. (2004) 'Theorizing food insecurity: building a livelihood vulnerability framework for researching food insecurity', *Norsk Geografisk Tidsskrift-Norwegian Journal of Geography*, 58(1): 23–37.

Yuan-Dong, Z., Xiao-He, Z. and Shi-Rong, L. (2011) 'Correlation analysis on normalized difference vegetation index (NDVI) of different vegetations and climatic factors in Southwest China', *Yingyong Shengtai Xuebao*, 22(2): 323–330.

Zhang, S., Li, Z., Lin, X. and Zhang, C. (2019) 'Assessment of climate change and associated vegetation cover change on watershed-scale runoff and sediment yield', *Water*, 11(7): 1373.

3

Land Cover and Land Use Change and Climate Variability: Evidence from Longitudinal Geographic Information System Data in Shashemene and Tehuledere

Kefyalew Sahle Kibret, Zerihun Mohammed, Lily Salloum Lindegaard, Ninna Nyberg Sørensen, Neil Webster and Dessalegn Rahmato

Introduction

In addition to land, labour, finance and knowledge (Fernando, 2022), intervention activities need dependable information, as do all development activities. Intervention activities aimed at addressing land cover change and climate issues are important in ensuring a sustainable future. There are several intervention strategies that can be employed to mitigate the impact of human activities on land cover change and climate. These strategies include reforestation, afforestation, conservation agriculture, agroforestry and sustainable forest management practices, among others. The success of these intervention measures depends on several factors such as community participation, effective policy frameworks, and research and development initiatives that provide innovative solutions to the challenges facing land use systems and their impacts on climate change. In many cases, information is not consistent and needs to be updated in emerging economies. The dynamics of land use and land cover (LULC) are one of the pieces of information needed for intervention activities in rural areas (SATPALDA, 2018; Abebe et al, 2021; Kuma et al, 2022). In such areas, LULC change might drive biodiversity loss, have a negative effect on the ecosystem services and functions of a forest area, and exacerbate climate change impacts. When it comes to reforestation, afforestation and rehabilitation, LULC can also be beneficial.

The LULC change in Ethiopia that has attracted the most attention is deforestation. The decrease of the forest cover exacerbates climate change. LULC change trends and their many underlying causes are spatially and temporally variable on both a global and a local scale. The expansion of agricultural and settlement land, population growth, the production of charcoal, the expansion of coffee plantations and poor land management may be some causes of LULC in Ethiopia. In certain areas, deforestation and reforestation also contribute to LULC changes (Alemu et al, 2015; Gebrelibanos and Assen, 2015; Kindu et al, 2015; Belete et al, 2021).

This chapter examines climate variability and LULC changes, including trees outside forests (TOF), an important but underrepresented resource in forest-poor nations. TOF refers to trees growing in nonforest areas like agricultural land, urban settings and along roads. The definition is comprehensive, covering various types of trees and locations, and the concept acknowledges their ecological and economic benefits. However, different definitions can exclude certain vegetation, leading to data variability and inconsistencies. Additionally, global assessments of TOF are limited, making it challenging to capture their full impact. There is a need to describe and comprehend the dynamics of trees and shrubs on rural and urban land, and their interaction with forest dynamics (Bellefontaine, 2002). A study in India shows that TOF, mainly growing on private land, are the main source of wood in the country for industry and domestic woodfuel (Pandey, 2008). To date, TOF status and characterization have been unknown due to their frequent omission from national forest resource assessments and the lack of readily accessible very high-resolution remotely sensed imagery (Thomas et al, 2021). As a result, off-forest tree management should be better understood, and forest, farm, pastoral and urban land should all be managed in an integrated and sustainable way (Bellefontaine, 2002). Therefore, it is of great importance to study changes in LULC in both forest and nonforest areas.

The study areas, the Shashemene and Tehuledere districts, are known for their cereal production, population and livestock production, but are susceptible to land degradation, changes in forest vegetation cover and severe soil erosion. Changes in LULC can lead to environmental degradation and loss of ecosystem services that humans depend on for survival. In addition, the region is characterized by vegetation practices at different levels. Changes in LULC are known to have significant impacts on biodiversity (woody species, medicinal plants, wildlife and so on). Additionally, community tree-planting activities constitute non-agricultural income for the community.

Comprehensive information on the drivers and causes of LULC is therefore necessary to build appropriate environmental regulations and land management approaches for each study area and beyond. However, data are lacking and little is known about the trends, magnitude and extent of LULC change involving TOF in the study landscape, or its drivers and impacts.

This chapter addresses this information gap. In addition, it provides an example of a methodology to assess LULC dynamics considering woody vegetation outside forests using a combination of remote sensing and geographic information systems (GIS) approaches. This is otherwise a gap in current research, limiting knowledge production on this topic. The main objective of the analysis was to perform LULC change detection using GIS and remote sensing, with specific objectives: (i) analysing trends in precipitation and temperature; (ii) determining LULC types at selected points in time; (iii) assessing the LULC changes over the past 34 years; and (iv) analysing the factors driving the changes. Background information of the study area was needed to support baseline data collection for monitoring and climate management. The chapter thereby documents long-term changes in vegetation (that is, change detection) and climate variables (that is, precipitation and temperature) in the two selected districts. These results can be used by land-use planners, ecologists, policy makers and other stakeholders to develop environmentally sound planning, policies and strategies.

Location

This study was conducted in two selected districts. The districts are located in north and central Ethiopia (Figure 3.1). Tehuledere is located at the eastern edge of the Ethiopian highlands in the Debub Wollo zone. The altitude of Tehuledere ranges from 500 metres above sea level along the boundary with the Debub Wollo zone to 2,700 metres along its southwest border. The hydrology of this *woreda* includes two lakes, Hayq and Ardibbo, which lies to the south of Hayq. Shashemene *woreda* is part of the West Arsi zone located in the Great Rift Valley. The altitude of this *woreda* ranges from 1,500 to 2,300 metres above sea level.

The climate in the study regions

Ethiopia is divided into 17 agro-ecological zones based on elevation and rainfall (Bekele-Tesemma et al, 1993; Bekele-Tesemma and Tengnäs, 2007; Hurni et al, 2016). The agro-ecological classification based on the elevation are Bereha (desert, < 500 metres), Kolla (lowlands, 500–1,500 metres), Weyna Dega (midlands, 1,500–2,300 metres), Dega (highlands, 2,300–3,200 metres), Wurch (highlands, 3,200–3,700 metres) and Kur (highlands, >3,700 metres). According to the elevation based agro-ecological classification, Tehuledere *woreda* lies in the Kolla and Weyna Dega agro-ecological zones.

Cereal crops in Ethiopia have two growing seasons. Any crop harvested between March and August is a 'Belg' season crop, while crops harvested

Figure 3.1: The two study districts and their location in Ethiopia

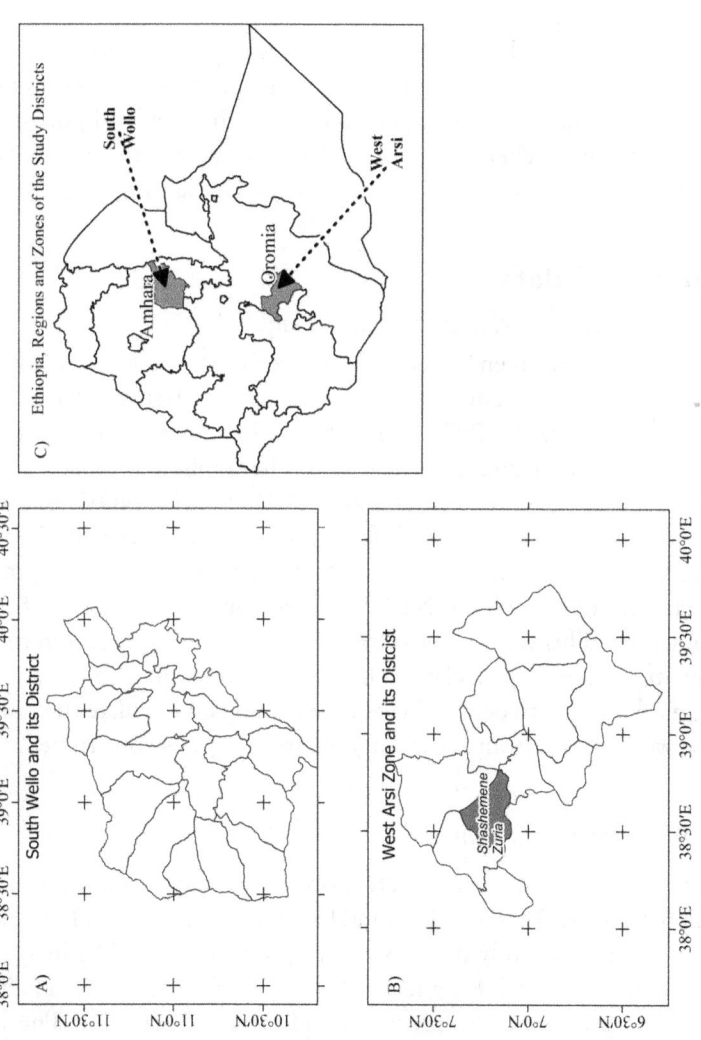

Source: Authors' construct, originally published open access in D. Rahmato, Z. Mohammed and N.A. Webster (2021) 'Governing climate mobility: a scoping study of two localities in Ethiopia'. Danish Institute for International Studies, DIIS Working Paper No. 13.

between September and February are 'Meher' season crops. Meher is the main cropping season (CSA, 2012). Tehuledere *woreda* has two seasons in a 'normal' year (Orr et al, 2020), that is, a year with average conditions (Figure 3.2). In the first wet season (Belg), smallholders may plant wheat, teff or long-duration sorghum. In the main wet season (Meher) they may plant early maturing sorghum, wheat or teff. The farmer's choice of cereal crop largely depends on rainfall.

Shashemene *woreda* lies in the Weyna Dega agro-ecological zone. The district has two seasons in a 'normal' year (Figure 3.3). In the first wet season (Belg), smallholders plant short period crops such as teff and potato. In the main wet season (Meher), they may plant wheat, potato or teff. Double cropping is widely practised provided that the rainfall is normal.

Methods and data

Approach to vegetation change detection analysis

The analysis of the green cover change over time was carried out in the selected districts using remote sensing data. The analysis was based on two points in time: 1986 and 2020. Landsat-5 and Landsat-8 images were used for 1986/1988 and 2020, respectively. The applied approach used the normalized difference vegetation index (NDVI). The NDVI can also be used to categorize and classify different land cover types and changes by the suitable threshold values. For each district, independently produced image classification results of NDVI images from the two dates of interest were obtained. This was followed by a pixel-by-pixel comparison to detect changes in the land cover classes. By adequately coding the classification results, we were able to define the change classes and calculate the transition rates between classes from the change matrix that was constructed.

Data for change detection analysis

The input data for the change detection analysis was made using Level-1 Landsat-5 Thematic Mapper (TM) and Landsat-8 (Operational Land Imager) images covering the study districts in two points in time. The images were accessed in Google Earth Engine (GEE)[1] (Gorelick et al, 2017). The images were already geometrically (precision and terrain) corrected. The images were visually interpreted and analysed to check the quality of the images. The images were selected from the dry period. January and December had the most cloud-free and usable images for the study sites. These months also coincided with the dry season where subtle rainfall effects on vegetative growth are unlikely; thus, they are the most suitable for conducting green cover rate and change level analysis spanning several years. Priority was given to anniversary and near-anniversary date images to reduce scene-to-scene

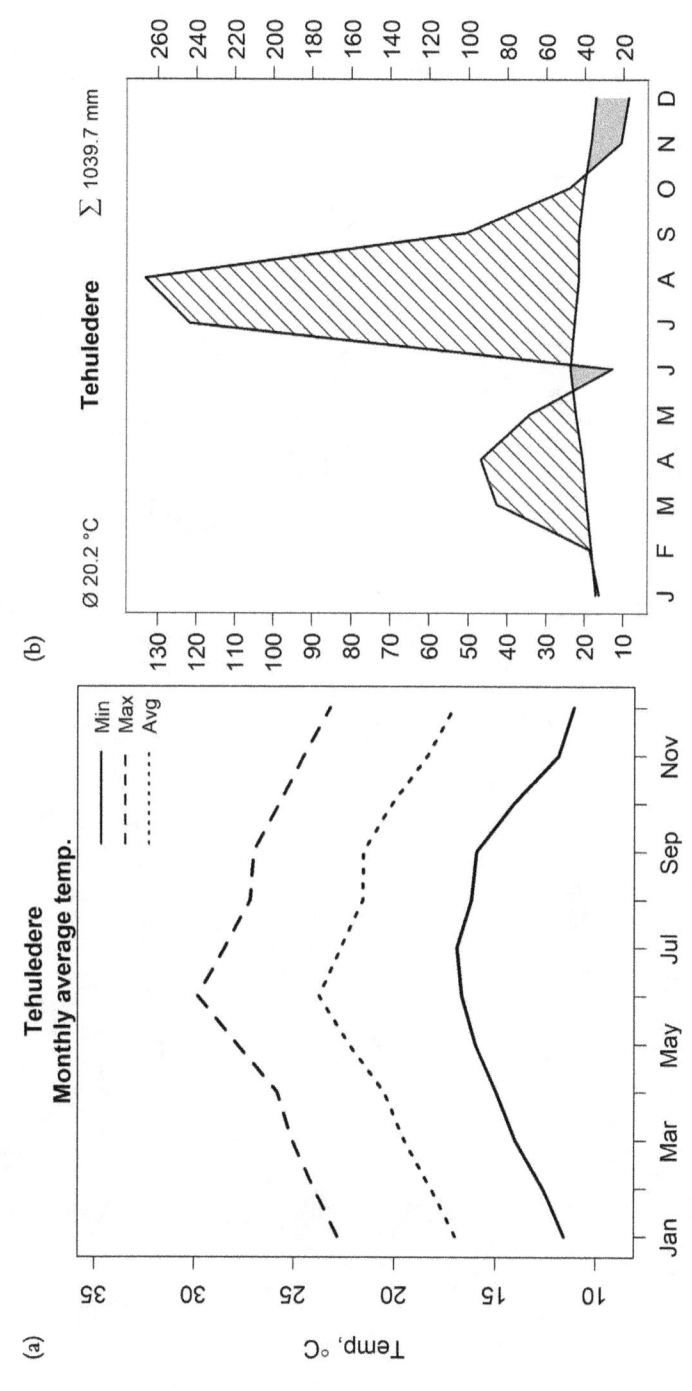

Figure 3.2: Long-term climatic changes in Tehuledere from 1982 to 2020 in (a) average monthly temperatures (minimum, average, and maximum) and (b) average monthly rainfall

Source: Authors' construct, originally published in Rahmato et al (2021)

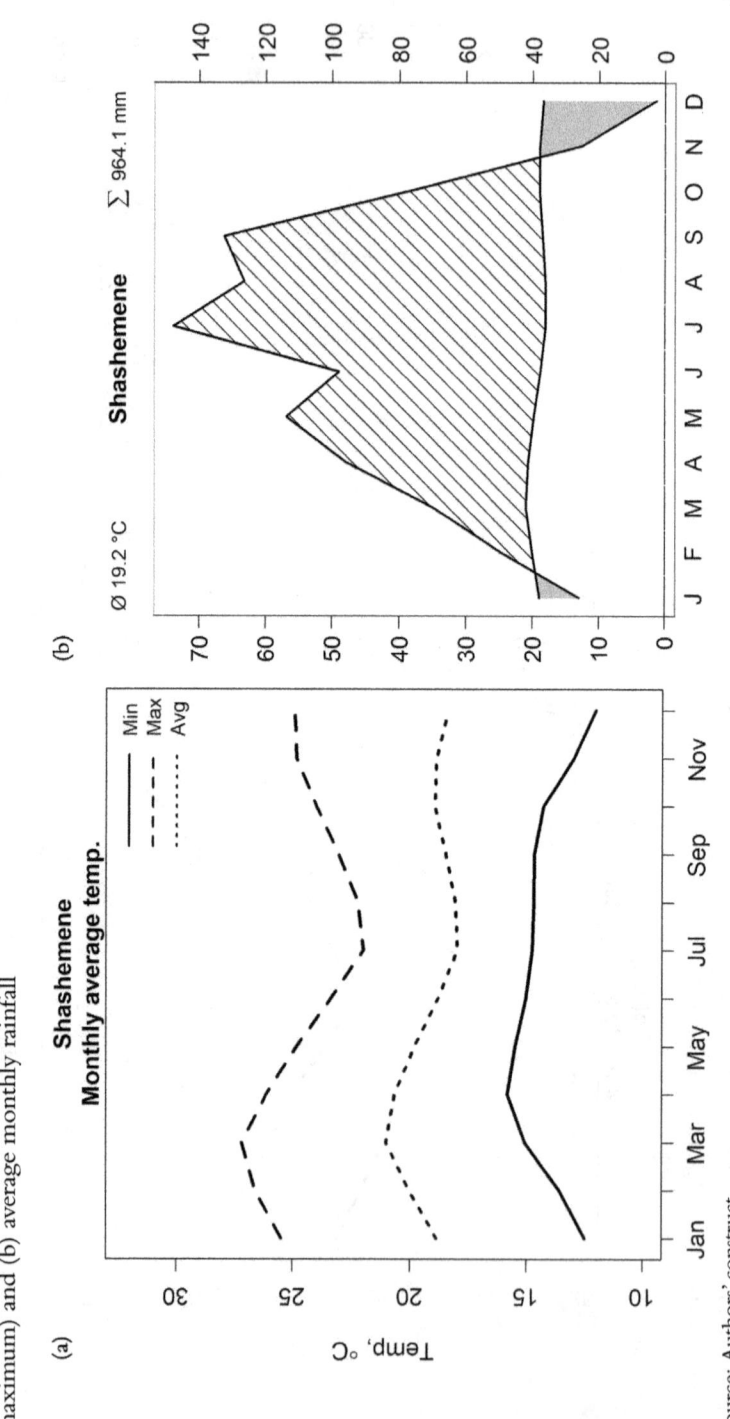

Figure 3.3: Long-term climatic changes in Shashemene from 1982 to 2020 in (a) average monthly temperatures (minimum, average, and maximum) and (b) average monthly rainfall

Source: Authors' construct

Table 3.1: Image datasets used for land cover change analysis, Ethiopia

District	Sensor	Bands	Path/Row	Date
Tehuledere	Landsat-5 – TM	1–7	168/052	5 January 1986
Tehuledere	Landsat-8 – OLI	1–9	168/052	5 January 2021
Shashemene	Landsat-5 – TM	1–7	168/055	21 January 1986
Shashemene	Landsat-8 – OLI	1–9	168/055	19 January 2020

Table 3.2: Satellite sensor characteristics, Ethiopia

Sensor	Characteristics		
	Temporal resolution	Earth–Sun distance	Swath width
Landsat TM	16 days	0.98	185km
Landsat OLI	16 days	0.98	185km

variation due to sun angle, soil moisture, atmospheric condition and vegetation phenology differences (Table 3.1). The characteristics of the different satellite sensors involved are shown in Table 3.2.

Data pre-processing

The georectification of the selected images was not required as these were Level-1 Precision and Terrain (L1TP) corrected data. The selected images were calibrated top-of-atmosphere (TOA) reflectance. Calibration coefficients were extracted from the image metadata in GEE (Chander et al, 2009).

Normalized difference vegetation index computation and image classification

The NDVI was used to quantify the green cover, as it is useful in understanding vegetation density and health and assessing changes over space and time. It is calculated as a ratio between the R and NIR bands and is given as:

$$NDVI = \frac{NIR - R}{NIR + R}$$

The NDVI images for all the datasets were computed in GEE environment. The NDVI images were subsequently classified using the threshold values provided by ESRI (2014) as shown in Table 3.3. Land cover classes for each

Table 3.3: Classification of the green cover type, Ethiopia

Code	NDVI range	Land cover types	Land cover types (mapped)	Label
6	< 0	Water	Water	W
1	0.0–0.3	Cropland/Grassland/Bare land/Built areas	Cropland/Grassland/Bare land/Built areas	CG
2	0.3–0.4	Grassland/cropland		
3	0.4–0.5	Shrub land/trees outside forest	Shrub land/trees outside forest/grass	S
4	0.5–0.6	Open forest/trees outside forest	Forest	F
5	≥ 0.6	Dense forest		

Table 3.4: Change matrix between the tree cover and nontree cover class from time 1 to time 2

Nr	Time 1	Time 2	Change
1	CG	S, F	Tree cover gain
2	S, F	CG	Tree cover loss
3	S, F	S to S, S to F, F to S, to F	Tree cover stable
4	CG, W	CG, W	Nontree cover stable

Note: For a description of Landcover codes, see Table 3.3.

independently produced image classification result were obtained. A change matrix was then constructed to define the change classes and to calculate the transition rates between classes for each pair of dates being analysed. The threshold interpretation should consider the local conditions.

To analyse the spatial distribution of the changes in vegetation cover between the two times, a woody cover changes analysis was done. Table 3.4 shows the change types. Four change types were considered from time 1 to time 2 with emphasis on tree and nontree covers. Tree cover in this case considered forest and shrub covers as one category and nontree cover (including water, cropland, grassland, bare land and built areas) as the other category. The four types of changes considered with respect to tree cover were: (1) tree cover gain; (2) tree cover loss; (3) stable tree cover; and (4) stable nontree cover.

To assess the spatial distribution of the tree cover change outside the forest cover, the shrub cover (TOF) was assessed during the two points of time, with emphasis on the TOF and/or the shrub cover change. In most cases, places classified as shrub were located in mountainous terrain and in the agricultural land use-dominated landscapes. Table 3.5 shows the types of changes with emphasis on shrub/TOF between time 1 and time 2.

Table 3.5: Change matrix between the shrub cover and nonshrub cover class from time 1 to time 2

Nr	Time 1	Time 2	Change
1	CG, F	S	Shrub cover gain
2	S	CG, F	Shrub cover lose
3	S	S	Shrub cover stable
4	CG, W,	CG, W, F	Nonshrub cover stable

Note: For a description of Landcover codes, see Table 3.3.

The aim of the results of the tree cover change and the shrub/TOF change analysis was to show the extent of the different change types and their spatial location.

Climate data analysis

Historical climate-related information was generated for each site using raw data available in different online archives that could be accessed via GEE (Gorelick et al, 2017). Climate Hazards Group InfraRed Precipitation with Station data (CHIRPS) was used as a source of data for rainfall analysis. CHIRPS is a 30+ year quasi-global rainfall dataset (Funk et al, 2015). CHIRPS incorporates 0.05° resolution satellite imagery with in situ station data to create a gridded rainfall time series for trend analysis and seasonal drought monitoring (Funk et al, 2015). ERA5-Land was used to derive temperature data for each site. It is a re-analysis dataset providing a consistent view of the evolution of land variables over several decades. Re-analysis combines model data with observations from across the world into a globally complete and consistent dataset using the laws of physics (Muñoz Sabater, 2019). It produces data that goes several decades back in time, providing an accurate description of the climate of the past. This dataset includes all 50 variables as available on Climate Data Store (CDS). Some of these are minimum, maximum and average temperature at 2 metres above ground level.

The climate-related analysis conducted for each site included rainfall and temperature trend analysis, rainfall anomaly analysis and coefficients of variation (CV) analysis.

The trend analysis was done using simple regression analysis for the rainfall and temperature using annual and seasonal datasets. The rainfall trend was also analysed using a nonparametric approach. Sen's slope, seasonal Sens's slope and the Mann-Kendall test were applied to the annual and monthly rainfall time series data (1982–2021). For linear trends, the slope is usually estimated by computing the least squares estimate using linear regression.

However, it is only valid when there is no serial correlation, and the method is very sensitive to outliers. A more robust method was developed by Sen (1968). Sen's slope estimator indicates slope (magnitude and direction) of any time series. Positive values are indicative of an increasing trend and vice versa (Sen, 1968). These values correspond to the rate of change (via slope) in the time series. Sen's slope is basically used to identify the magnitude of the trend in a data series which is not serially auto-correlated, and the Mann-Kendall test is used to identify the significance level (determining if the magnitude is statistically significant or not) at different confidence levels (95 or 99 per cent). Sen's slope computes Sen's slope for linear rate of change and corresponding confidence intervals. Seasonal Sen's slope computes seasonal Sen's slope for a linear rate of change. The Mann-Kendall trend test is used to determine whether a time series has a monotonic upward or downward trend. It does not require that the data be normally distributed or linear.

The CV is a standardized measure of the dispersion of a probability distribution or frequency distribution. It is defined as the ratio of the standard deviation to the mean, and is often expressed as a percentage. The CV in this study was computed for the rainfall of each site. The computation was done using monthly, seasonal and annual datasets. The results are presented as a table for simple comparison, while the results of the climate analysis are presented as a chart and table.

Rainfall and temperature trend

The long term annual and seasonal rainfall indicated no significant change over the last two decades (1982 and 2020) in the Tehuledere district. Both the long-term annual and seasonal rainfall (except for the main rainy season) indicated neither an increase nor a decrease over the last four decades (1982 and 2020) (Figure 3.4). The main rainy season in July, August and September (referred to here as JAS) indicated a very slight increase in rainfall ($R^2 = 0.14$ and $P < 0.05$).

The long-term maximum temperature showed an increasing trend for both annual and rainy seasons in Shashemene (Figure 3.5). The increase in maximum temperature was strong for the dry season of January and February. The increase in temperature during the main rainy season would have a substantial impact on increased evapotranspiration and swift dry-out of soil moisture.

The long-term maximum temperature showed an increasing trend for dry seasons in Tehuledere (Figure 3.6). The maximum temperature increase was strong for the season from March to May.

The outputs of the Mann-Kendall trend test and Sen's slope test were used to test the annual and monthly rainfall change over time for the Shashemene district. Outputs of the Mann-Kendall trend test analysis of the annual

Figure 3.4: Rainfall trend from 1982 to 2020 (annual) and seasonal (January to February [JF], March to May [MAM] and July to September [JAS]) in Tehuledere

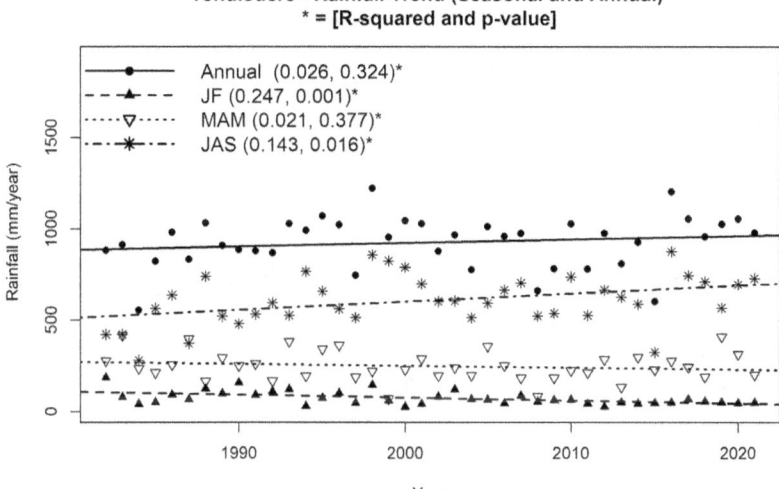

Source: Authors' construct

Figure 3.5: Temperature trend from 1982 to 2020 (annual) and seasonal (January to February [JF], March to June [MAMJ] and July to October [JASO]) in Shashemene

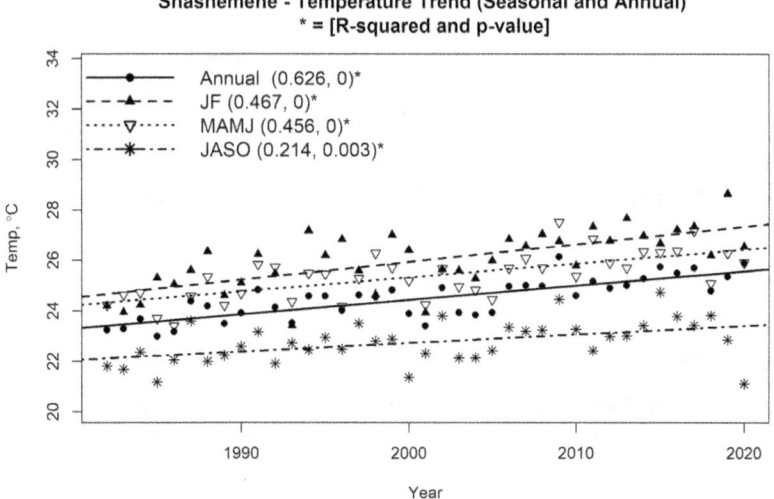

Source: Authors' construct

Figure 3.6: Temperature trend from 1982 to 2020 (annual) and seasonal (January to February [JF], March to May [MAM] and July to September [JAS]) in Tehuledere

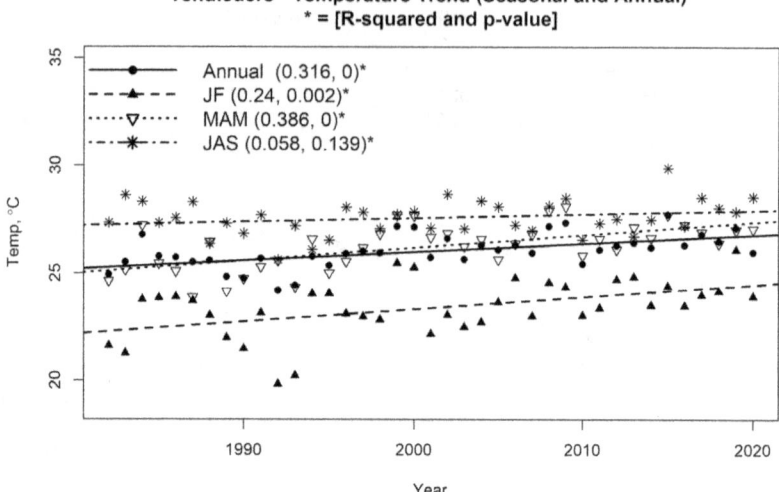

Source: Authors' construct

rainfall time-series data indicated that annual rainfall has not changed since 1982, with a 5 per cent significance level (z = 0.198068, n = 40 and p-value = 0.842992), whereas the seasonal Sen's slope was estimated to be 0.065 for the seasonal rainfall data (40 years × 12 months). Similarly, the outputs of both tests were used to test the annual and monthly rainfall change over time for the Tehuledere district. Outputs of the Mann-Kendall trend test analysis of the annual rainfall time-series data indicated that annual rainfall has not changed since 1982, with a 5 per cent significance level (z = 0.66411, n = 40 and p-value = 0.506621), whereas the seasonal Sen's slope was estimated to be 0.018 for the seasonal rainfall data (40 years × 12 months). Generally, the seasonal analysis results show that there is a statistically significant increasing trend in the seasonal rainfall data in both districts.

Climate hazards

The annual and seasonal climate hazards were identified by calculating the rainfall anomaly. Rainfall anomaly is used to compare the long-term variation for each individual year or seasons. This gives an estimate of the degree to which rainfall recorded during the year or season is either stable or changing. A negative anomaly indicates that the year is characterized by dry climatic conditions in the season in question, while the positive anomaly shows it is characterized by wet climatic conditions. We employ the index classification by Agnew and Chappell (1999), of extreme drought ($z < -1.65$), severe drought ($-1.28 > z > -1.65$), moderate drought ($-0.84 > z > -1.28$) and no drought ($z > -0.84$). The annual rainfall anomalies at Shashemene range from above -2.0 in 1984 and 2002, indicating extreme drought, to above +2 in 1996, indicating no drought (wettest) (Figure 3.7a). The seasonal anomaly analysis indicated variability in the annual rainfall occurrence of moisture content with both positive and negative anomalies during the analysis period (Figure 3.7b, c and d). The occurrence of moisture deficiency was severe during the first rainy season (March to June) in 1984 and 1988 (Figure 3.7c); during the main rainy season (July to October), it was severe in 1984, 1986, 2002, 2009 and 2016 (Figure 3.7d).

The annual rainfall anomaly at Tehuledere ranges from above -2.0 in 1984 and 2015, indicating extreme drought, to above +2.0 in 1998 and 2018, indicating no drought (wettest) (Figure 3.8a). The seasonal anomaly analysis indicated variability in the annual rainfall occurrence of moisture content with both positive and negative anomalies during the analysis period (Figure 3.8a and d). The occurrence of moisture deficiency was severe during the first rainy season (March to May) in 1999 and 2013 (Figure 3.8c); during the main rainy season (July to September), it was severe in 1982, 1983, 1984, 1987 and 2015 (Figure 3.8d).

The level of mean rainfall need not necessarily be a constraint on successfully carrying out agriculture in this area. However, mean annual rainfall does not indicate the natural year-to-year variability of rainfall that occurs in the observed area. As a result, the coefficient of variation (CV expressed as a percentage) was computed based on the monthly, seasonal and annual rainfall dataset of the area. In general, the higher the CV, the more variable the year-to-year (that is, inter-annual) rainfall of a locality. The Tehuledere district received about 65 per cent of its annual rainfall from July to September and one quarter of its rainfall from March to May. The Shashemene district received about 52 per cent of its annual rainfall during the main growing period (July to October) and 40 per cent during the short rainy season (March to June) (Table 3.6).

According to Hare (1993), CV (per cent) values are classified as: < 20 per cent as less variable, 20–30 per cent as moderately variable and > 30 per cent as highly variable. In Tehuledere, the intermonthly CV for the main rainy season from March to September ranged between 25 and 73 per cent – that is, between moderately to highly variable (Table 3.7). The main rainy season is characterized by moderate and highly variable CV (25–36 per cent). In Shashemene, the intermonthly CV for the main rainy season from March to October ranged between 16 and 55 per cent, which is between low and highly variable (Table 3.7). The first rainy season is characterized by a moderate to highly variable CV (23–66 per cent). The

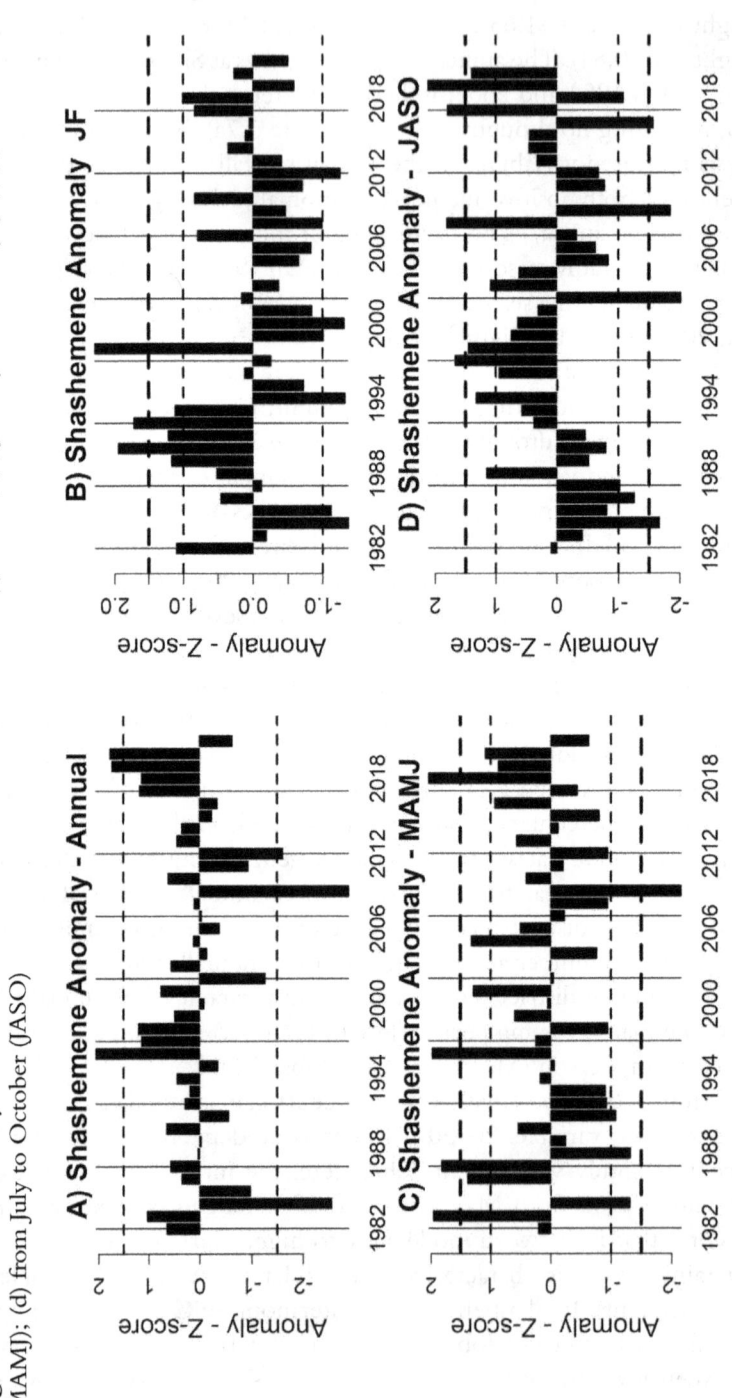

Figure 3.7: Rainfall anomaly from 1982 to 2021 in the Shashemene district: (a) annual; (b) January to February (JF); (c) from March to June (MAMJ); (d) from July to October (JASO)

Source: Authors' construct

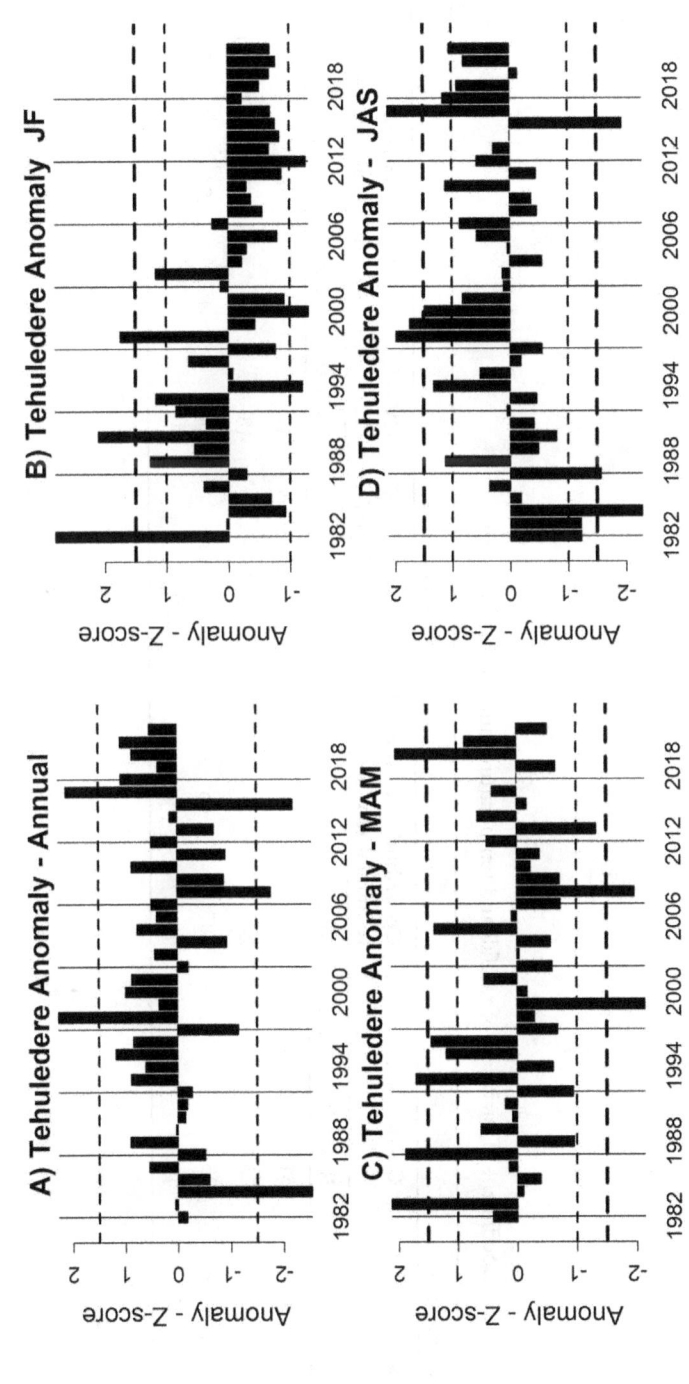

Figure 3.8: Rainfall anomaly from 1982 to 2021 in the Tehuledere district: (a) annual; (b) January to February (JF); (c) from March to May (MAM); (d) from July to September (JAS)

Source: Authors' construct

Table 3.6: Coefficient of variation of rainfall and temperature (annual and by season)

Tehuledere					Shashemene					
Season	CV of		Contribution of rainfall to annual rainfall		Season	CV of		Contribution of rainfall to annual rainfall		
	Rainfall	Temp.	mm	%		Rainfall	Temp.	mm	%	
January–February	51	6	73	8	January–February	49	5	76	8	
March–May	33	4	248	27	March–June	18	4	378	40	
July–September	23	3	607	65	July–October	14	4	484	52	
Annual	15	3	927	100	Annual	11	3	938	100	

Table 3.7: Inter-annual coefficient of variation, monthly rainfall contribution to the annual rainfall (mm = monthly rainfall and % = monthly contribution in %)

Month	Tehuledere			Shashemene		
	mm	CV	%	mm	CV	%
Jan	33	59	3	26	56	2
Feb	39	71	3	50	66	5
Mar	85	56	8	70	44	7
Apr	93	43	8	96	38	9
May	68	62	6	113	30	11
Jun	25	73	2	97	23	10
Jul	240	36	23	147	21	15
Aug	265	25	25	127	16	13
Sep	100	29	9	131	22	13
Oct	49	77	4	76	55	7
Nov	21	82	2	25	67	2
Dec	17	66	1	2	153	0

main rainy season is characterized by a low to highly variable CV (16–55 per cent). Therefore, the major problem facing the study area was rainfall variability between months, which is linked to either late onset or early offset in rainfall. In Shashemene, the amount and timing of rainfall in March to May in particular, which accounts for 7–11 per cent of the annual rainfall, are vital for the establishment of crops sown during this time.

Land cover change

Land cover changes in the Tehuledere district between 1986 and 2021

Figure 3.9 shows the result of the image classification for the Tehuledere district for various land cover classes for January 1986 and January 2021, respectively. Table 3.8 shows detailed statistics of the various land cover classes for the two years under analysis. Figure 3.9a shows the 1986 land cover situation. The area covered by crop/grassland accounted for 38.03 km² (85.3 per cent) and shrubland for 23.23 km² (5.2 per cent). The district was covered by 4.94 km² of forest vegetation, which is about 1 per cent of the total area of 445.67 km². This was mainly found in the southern part of the district. The area covered by water was estimated to be 36.91 km² (8.3 per cent). Figure 3.9b shows the land cover situation in 2021. The area covered by crop/grass decreased to 231.96 km² (52 per cent) from 1986 to 2021. The forest and shrub cover increased to 87.16 km² (19.56 per cent) and 85.92

Figure 3.9: Land cover changes in the Tehuledere district between (a) January 1986 and (b) January 2021

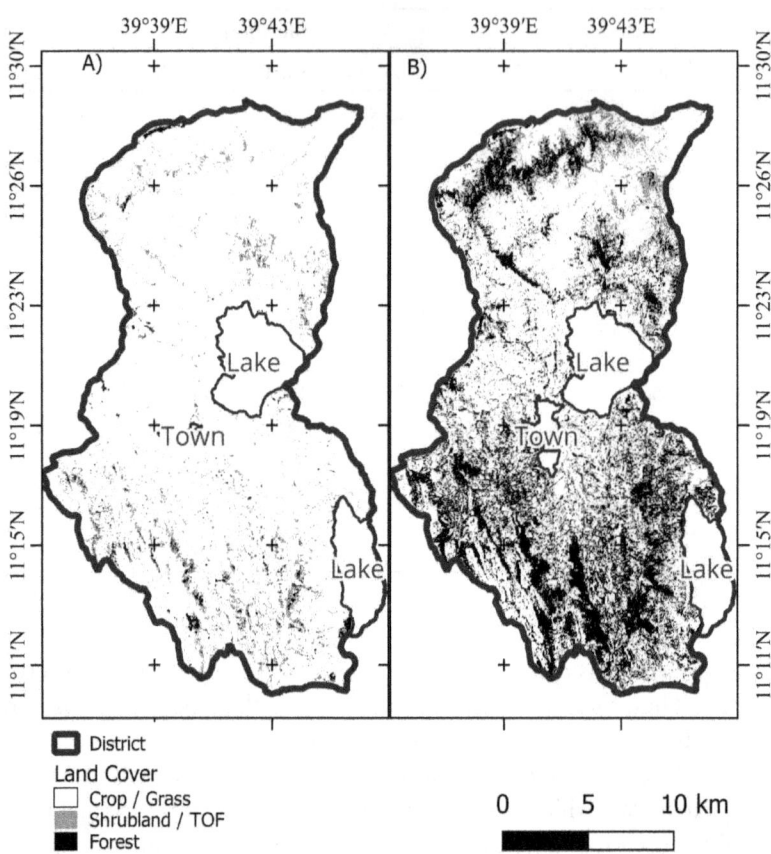

Source: Authors' construct

km² (19.30 per cent), respectively, from 1986 to 2021. The area covered by water decreased to 36.37 km² (8.16 per cent) during the same period. The change in land cover from 1986 to 2020 should be interpreted with care. The dark grey colour in Figure 3.9b is not necessarily forest; instead, it can be a mosaic of cropland, grass and woody vegetation if it is located in the agricultural land-use dominated landscape. Most of the forests are found in the mountainous areas.

The different land cover types presented previously (the maps and table) do not show the spatially explicit change patterns. Table 3.9 shows the spatially explicit change types with an emphasis on woody vegetation cover from 1986 to 2021. Figure 3.10 shows the woody vegetation cover gained from 1986 to 2021, which was estimated to be 148 km² (33.6 per cent of the total area). The woody vegetation cover lost was estimated to be 3 km² (1 per

Table 3.8: Land cover changes in the Tehuledere district between January 1986 and 2021

Land cover	1986			2021			Change (+/-)	
	Pixels	Area (km²)	%	Pixels	Area (km²)	%	Area (km²)	%
Water	41,008	36.91	8.3	40,416	36.37	8.16	-0.54	-1.5
Forest	5,494	4.94	1.1	96,840	87.16	19.56	82.22	1,664.4
Shrub	25,807	23.23	5.2	95,472	85.92	19.30	62.69	269.9
Crop and grass	422,257	380.03	85.3	257,730	231.96	52.00	-148.07	-39
Town	624	0.56	0.1	4,732	4.26	1.00	3.7	660.7
Total	495,190	445.67	100	495,190	445.67	100		

Table 3.9: Woody vegetation (forest/shrub land) cover change from 1986 to 2021 in the Tehuledere district

Change type	Pixels	Percentage	Area (km²)
Gained woody vegetation	164,568	33	148
Lost woody vegetation	3,776	1	3
Stable woody vegetation	27,518	6	25
Stable nonwoody vegetation	299,328	60	269
Total	495,190	100	446

cent) and occurred in the southeastern part of the district. The 1986 forest which was still forest in 2021 was 25 km² (6 per cent). This is represented as stable woody vegetation cover area and it is marked as black on the map. The fourth type of change is the stable nonwoody vegetation cover, which is represented as white in the map and makes up the largest share.

The detailed analysis of the woody vegetation cover shows that the change in the shrub/TOF cover was mainly in the agricultural land use-dominated landscape and around the forest land. Figure 3.10b shows the spatial distribution of the TOF that were gained during 1986 to 2021. The gained shrub/TOF accounted for 82 km², that is, 18 per cent of the total area of the district (Table 3.10). During the same period, only 2 km² of shrub/TOF were lost. Shrub land/TOF that did not show change during the same period was estimated to cover an area of 4 km² (1 per cent). The gained and stable shrub/TOF were mainly located in settlement areas and long streams. Most of these were woodlots of eucalyptus.

Land cover changes in the Shashemene district between 1986 and 2020

Figure 3.11 shows the result of the image classification for various land cover classes for the Shashemene district for January 1986 and January 2020. Table 3.11 shows detailed statistics of the various land cover classes for the two years under analysis. Figure 3.11a shows the 1986 land cover situation. The area covered by crop/grass accounted for 630 km² (88 per cent). The shrub and forest cover were estimated to be 34 km² (4.9 per cent) and 34.98 km² (4.5 per cent), respectively. The area covered by water was estimated to be 1.71 km² (0.2 per cent). Figure 3.1b shows the land cover situation in 2020. The cropland decreased to 529.4 km² (74 per cent) from 1986 to 2020. The forest cover increased to 82.59 km² (11.5 per cent) in the same period, whereas the area covered by shrubs increased to 59.57 km² (8.3 per cent). The area covered by water decreased to 0.79 km² (a mere 0.1 per cent). The decline of cropland is attributed to the expansion of the Shashemene

Figure 3.10: Spatial distribution of changes: (a) woody vegetation cover change; and (b) shrub/trees outside forests in the Tehuledere district

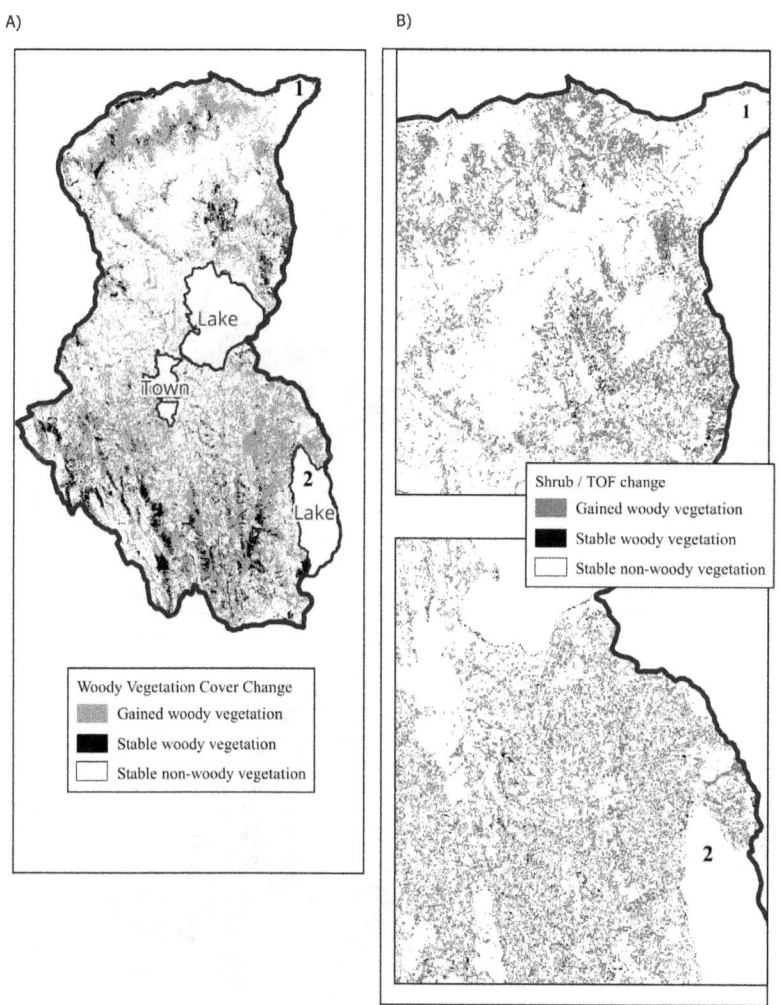

Source: Authors' construct

Table 3.10: Shrub/trees outside forests change from 1986 to 2020 in the Tehuledere district

Change type	Pixels	Percentage	Area (km²)
Gained shrub/TOF	90,616	18	82
Lost shrub/TOF	2,757	1	2
Stable shrub/TOF	4,737	1	4
Stable nonshrub/TOF	397,080	80	357
Total	495,190	100	446

Figure 3.11: Land cover changes in the Shashemene district between (a) 1986 and (b) 2020

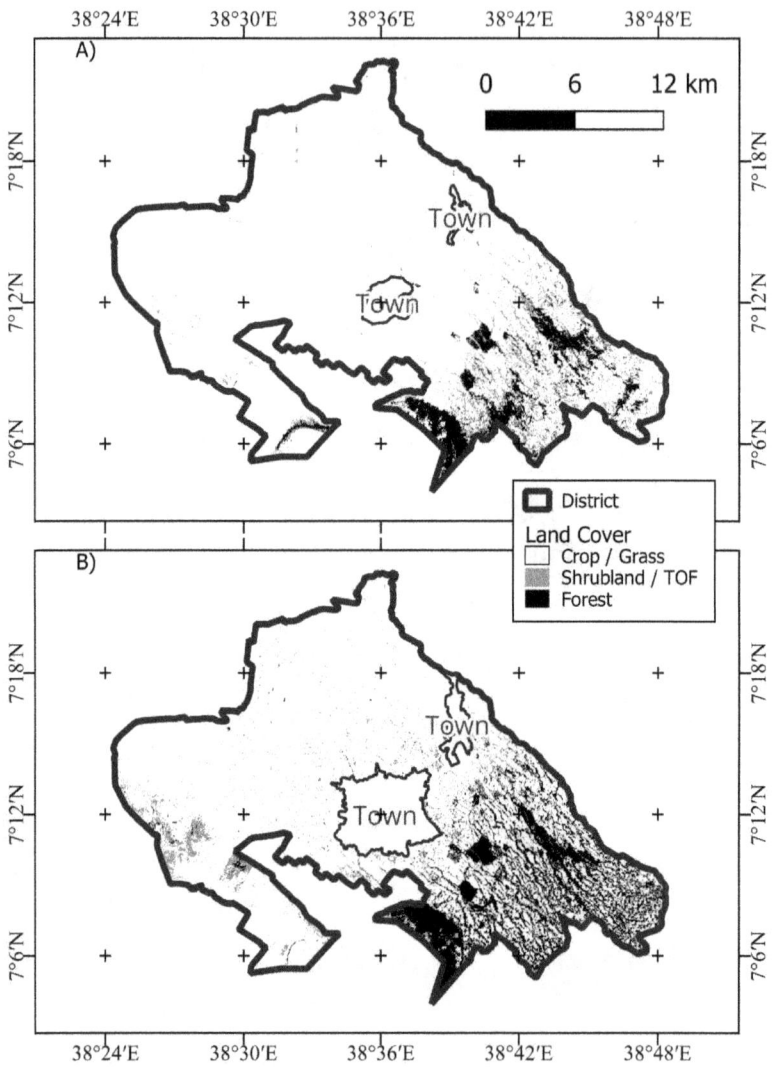

Source: Authors' construct

town, which came at the cost of the cropland. Cropland was also converted to woodland during the period. This is mainly found in areas where the land is managed by government institutions.

The information on different land cover types presented previously (Figure 3.11 and Table 3.11) do not show the spatially explicit change patterns. Figure 3.12 shows the spatially explicit change types with an

Table 3.11: Land cover changes in the Shashemene district between January 1986 and January 2020

Land Cover	1986			2020			Change (+/-)	
	Pixels	Area (km²)	%	Pixels	Area (km²)	%	Area (km²)	%
Water	1,899	1.71	0.2	878	0.79	0.1	-0.92	-53.8
Forest	38,869	34.98	4.9	91,770	82.59	11.5	47.61	136.1
Shrub	38,653	34.79	4.9	66,186	59.57	8.3	24.78	71.2
Crop and grass	700,154	630.14	88	588,226	529.4	74	-100.74	-16
Town	16,105	14.49	2	48,620	43.76	6.1	29.27	202
Total	795,680	716.11	100	795,680	716.11	100		

Figure 3.12: Spatial distribution of changes: (a) woody vegetation cover change; and (b) shrub/trees outside forests cover change from 1986 to 2020 in the Shashemene district

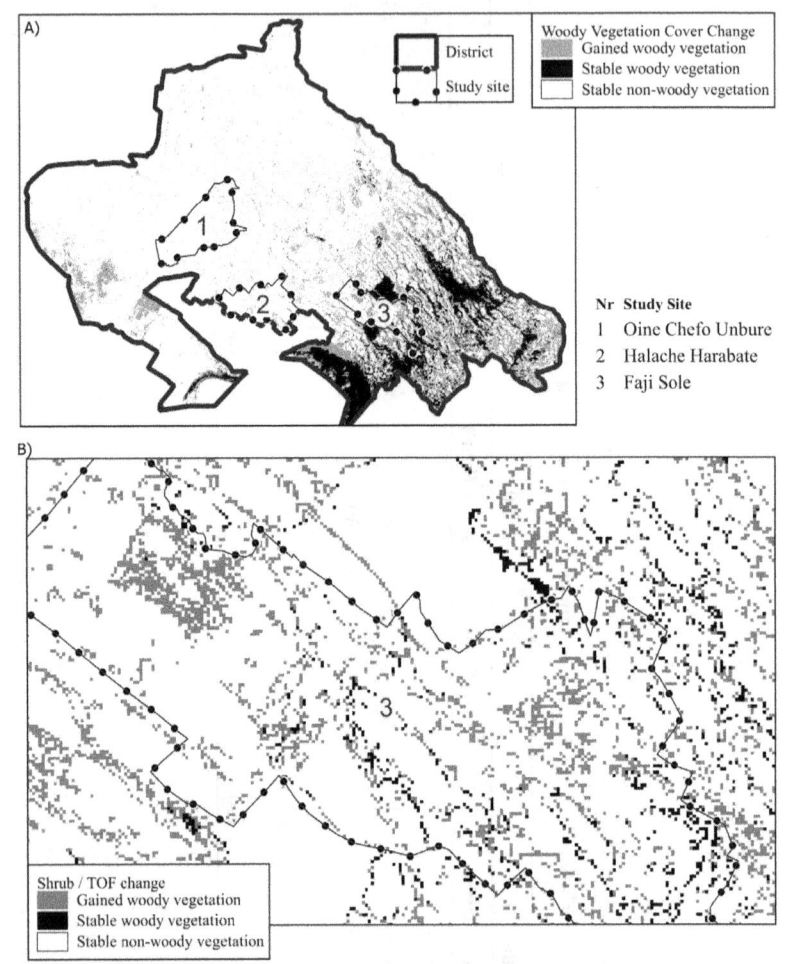

Source: Authors' construct

emphasis on woody vegetation cover from 1986 to 2020. Figure 3.12a shows the woody vegetation cover gained from 1986 to 2020, which was estimated to be 91 km² (13 per cent of the total area) (Table 3.12). This change occurred mainly in the agricultural dominated landscape and along the forest areas. The woody vegetation cover lost was estimated to be 18 km² (3 per cent) and occurred in the southeastern part of the district. The majority of the 1986 woody vegetation cover remained woody vegetation in 2020. This is represented as the stable woody vegetation cover area and covered 52 km² (7 per cent), and it was marked as black on the map. The

Table 3.12: Woody vegetation (forest/shrub) cover change from 1986 to 2020 in the Shashemene district

Change type	Pixels	Percentage	Area (km^2)
Gained woody vegetation	100,597	13	91
Lost woody vegetation	20,161	3	18
Stable woody vegetation	57,359	7	52
Stable nonwoody vegetation	617,563	78	556
Total	795,680	100	716

Table 3.13: Shrub/trees outside forests cover change from 1986 to 2020 in the Shashemene district

Change type	Pixels	Percentage	Area (km^2)
Gained shrub/TOF	55,104	7	50
Lost shrub/TOF	12,514	2	11
Stable shrub/TOF	11,082	1	10
Stable nonshrub/TOF	716,980	90	645
Total	795,680	100	716

fourth type of change is the stable nonwoody vegetation cover, which is represented as white on the map and makes up the largest share.

The detailed analysis of the woody vegetation cover shows that the change in the shrub/TOF cover was mainly in the agricultural land use-dominated landscape and around the forest land. Figure 3.12b shows the spatial distribution of the TOF that were gained during the period from 1986 to 2020. The gained shrub/TOF accounted for 50 km^2 (7 per cent of the total area of the district) (Table 3.13). During the same period, about 11 km^2 of shrub/TOF were lost. Shrub land/TOF that did not show any change during the same period was estimated to cover an area of 10 km^2 (1 per cent). The gained and stable shrub/TOF were mainly located in settlement areas and by a long stream. Most of these were woodlots of eucalyptus.

Discussion

Land degradation versus greening: assumptions versus reality and what this means for agrarian livelihoods and policy

Land-cover change is evident in both study areas and may have positive and negative consequences (Pellikka et al, 2013). Urban expansion was higher in the Shashemene district than in Tehuledere. In the Shashemene district, the expansion of the town has come at the cost of agricultural land use. This has been more pronounced in recent years. In both districts, on the positive side, agricultural expansion may increase food production for a growing population. However, the expansion of agricultural land has come at the cost of forest, shrubland and grassland in the Shashemene district, whereas in the Tehuledere district, grassland and shrubland were converted to agricultural land use. The second example of land cover conversion is the shrubland/TOF expansion in both districts. Woody vegetation cover has increased since 1986 in settlements and along streams. The woody vegetation in the Shashemene district is mainly eucalyptus trees.

In the Tehuledere district, an expansion of forest and shrubland was observed in the last three decades. Soil and water conservation is one of the most important benefits derived from community and state forests in the Tehuledere district. Both community and state forests located on the hillsides in the district provide soil and water conservation, and serve as natural habitats for wild animals (Woldie and Tadesse, 2019). Some participants in Woldie and Tadesse's study even cited examples of appreciably changed land cover and consequently a reduction of environmental degradation, gullies healed, springs developed, grass cover that has visibly increased on what was once bare land and an increase in the availability of wood products, because most of the land was formerly in a degraded condition. This land rehabilitation has resulted in improved availability of fodder, fuelwood and construction wood.

An example of more negative consequences of woody cover change is the decrease in soil water-holding capacity. As natural vegetation is replaced by agriculture, soil porosity may be reduced by soil compaction, decreasing infiltration capacity and increasing the risks of soil erosion (Pellikka et al, 2013). In the Shashemene district, there are areas that are affected by erosion. On the other hand, there are also areas that are affected by sedimentation and flooding. These areas are found in the lower part of the district. As soil water-holding capacity is reduced, the risk of hydrological droughts during dry seasons is increased, while during the rainy seasons, soils are more susceptible to erosion. Such soil loss and sediment-deposition processes may have a significant impact on agriculture, local economies and ecosystems (Alcántara-Ayala et al, 2006).

The land-cover changes taking place in the Shashemene district have been continuing since 1984 due to an ever-growing population and demand for land for cultivation of food crops. Land is evidently becoming a valuable natural resource in the area. Converting natural vegetation, forest or grassland to agricultural areas decreases biodiversity, reduces the capability of vegetation to capture atmospheric moisture and retain water in the vegetation cover, and exposes land to the risk of water erosion.

Change in land cover may also affect local microclimates; however, climate variability is influenced by a multitude of factors beyond the changes in land cover within the studied districts. These factors include atmospheric conditions, oceanic patterns and natural climate cycles, which all contribute to the overall variability observed. Therefore, isolating land cover change as a singular cause would be an oversimplification of the complex interactions affecting climate variability.

The role of alternative/additional income-generating activities in relation to climate change adaptation as well as migration

The study results reveal that on the one hand, access to the natural forest in the study areas has decreased, whereas on the other hand, off-forest tree coverage has increased in both districts. Likewise, vegetation coverage in farmers' homesteads has been increasing in both districts. Informal discussion with farmers in the Shashemene district indicated that trees are harvested for the construction market. The decline in natural forest and increase in demand for wood motivated people to privately plant trees. Tree planting is often recommended for farmers, but the extent of its encouragement depends on their socioeconomic status and practical considerations. Promoting private tree plantation can be considered as economic support and can fill the demand gap for fuelwood (Worku et al, 2017).

Although land available for plantation establishment is limited, tree coverage of the districts can also be increased by supporting technically appropriate species and variety selection and techniques of tree planting. In areas where suitable land for tree planting exists, local policy options for promoting allocation of unused land and mountainous topography for community and private tree planting, especially for landless and smallholder farmers, should be encouraged (Worku et al, 2017; Yinebeb et al, 2022). Concerning cultural importance, the cultural use of particular species is central to conserving plant species in household gardens in the study areas. Therefore, conservation plans should consider the determinants of floristic composition as well as the relationships between floristic composition and cultural importance. One study found that home-gardening involves a combination of vegetables, fruits and trees used for remedies, animal fodder, human consumption, household and farming tool fabrication, spice, house cleaning and recreational value for rural families (Akale et al, 2019). According to this study, home gardens, referring to small-scale production located by peoples' homes, might not be adequate to meet future needs. The findings presented in this chapter suggest that home gardens, including TOF, can serve future sustainable food security and land rehabilitation programmes. In future studies, such home-garden practices could be extended to the surrounding ecosystem through rehabilitation programmes.

Conclusions

The study was conducted in two districts in Ethiopia with the aim of examining the relationship between land cover change and climate change impacts. Remote sensing images were used to analyse changes in green cover over a 35-year period, including the examination of rainfall and temperature trends, as well as anomalies and variations. In Tehuledere district, no significant change in rainfall was observed over the last two decades, except for a slight increase during the main rainy season. However, there was an increasing trend in maximum temperatures, particularly during the dry season. In Shashemene district, there was no notable change in rainfall over the last four decades, except for a slight increase during the main rainy season. Maximum temperatures showed an increasing trend, mainly during the dry seasons. Overall, the analysis revealed some variability in rainfall occurrences, including severe moisture deficiency during certain seasons and years.

The variability of rainfall in a region is influenced by its CV, with a higher CV indicating more year-to-year fluctuation. In the Tehuledere district, 65 per cent of the annual rainfall occurs from July to September, while only 25 per cent occurs from March to May. Similarly, in the Shashemene district, 52 per cent of rainfall occurs during the main growing period (July to October) and 40 per cent during the short rainy season (March to June). The intermonthly CV for the main rainy season in Tehuledere ranges from moderately to highly variable (25–73 per cent), while in Shashemene, it ranges from low to highly variable (16–55 per cent). The main issue in the study areas is the variability between months, which can lead to late or early rainfall onset. In Shashemene, the amount and timing of rainfall in March to May are crucial for crop establishment. Additionally, the increase in forest cover can sometimes consist of patchy woody vegetation mixed with agricultural land rather than pure forests, which are mainly found in mountainous areas.

The analysis of woody vegetation cover reveals changes in shrub and tree cover, particularly in agricultural and forested areas. These changes were mainly observed in settlement areas and along streams, where eucalyptus woodlots were prevalent. Additionally, the expansion of Shashemene town resulted in a decline in cropland, with some areas being converted into woodland. This conversion was more prominent in areas managed by government institutions. Overall, the study highlights a growing trend in tree cover within the landscape.

Note

[1] https://code.earthengine.google.com/

References

Abebe, G., Getachew, D. and Ewunetu, A. (2021) 'Analysing land use/land cover changes and its dynamics using remote sensing and GIS in Gubalafito district, Northeastern Ethiopia', *SN Applied Science*, 4(1): 1–30.

Agnew, C.T. and Chappell, A. (1999) 'Drought in the Sahel', *GeoJournal*, 48(4): 299–311.

Akale, A.H., Alemu, M.V. and Asmamaw, M.B. (2019) 'Homegarden plants in Legambo District (Chiro Kebele) South Wollo, Ethiopia: future implication for food security and rehabilitation program', *African Journal of Plant Science*, 13(9): 246–254.

Alcántara-Ayala, I., Esteban-Chávez, O. and Parrot, J.F. (2006) 'Landsliding related to land-cover change: a diachronic analysis of hillslope instability distribution in the Sierra Norte, Puebla, Mexico', *CATENA*, 65: 152–165.

Alemu, B., Garedew, E., Eshetu, Z. and Kassa, H. (2015) 'Land use and land cover changes and associated driving forces in north-western lowlands of Ethiopia', *International Research Journal of Agricultural Science and Soil Science*, 5(1): 28–44.

Bekele-Tesemma, A. (2007) *Useful Trees and Shrubs of Ethiopia: Identification, Propagation, and Management for 17 Agroclimatic Zones*. Nairobi: RELMA in ICRAF Project.

Bekele-Tesemma, A., Birnie, A. and Tengnäs, B. (1993) *Useful Trees and Shrubs for Ethiopia: Identification, Propagation, and Management for Agricultural and Pastoral Communities*. Nairobi: Regional Soil Conservation Unit, Swedish International Development Authority.

Belete, F., Maryo, M. and Teka, A. (2021) 'Land use/land cover dynamics and perception of the local communities in Bita district, south-western Ethiopia', *International Journal of River Basin Management*, 21: 1–12.

Bellefontaine, R. (ed.) (2002) *Trees outside Forests: Towards Better Awareness, FAO Conservation Guide*. Rome: International Cooperation Centre on Agrarian Research for Development and Food and Agriculture Organization of the United Nations.

Chander, G., Markham, B.L. and Helder, D.L. (2009) 'Summary of current radiometric calibration coefficients for Landsat MSS, TM, ETM+, and EO-1 ALI sensors', *Remote Sensing of Environment*, 113(5): 893–903.

CSA (2012) *Agricultural Sample Survey 2011/2012 (September – December 2011) Report on Area and Production of Major Crops (Private Peasant Holdings, Meher Season)*. Statistical bulletin. Addis Ababa: Federal Democratic Republic of Ethiopia.

ESRI (2014) *ArcGIS 10.3 Help: NDVI Function* Redlands: ESRI Press.

Fernando, J. (2022) '4 factors of production explained with examples', *Investopedia*. Available from: https://www.investopedia.com/terms/f/factors-production.asp

Funk, C. et al (2015) 'The climate hazards infrared precipitation with stations: a new environmental record for monitoring extremes', *Scientific Data*, 2: 150066.

Gebrelibanos, T. and Assen, M. (2015) 'Land use/land cover dynamics and their driving forces in the Hirmi watershed and its adjacent agro-ecosystem, highlands of Northern Ethiopia', *Journal of Land Use Science*, 10(1): 81–94.

Gorelick, N. et al (2017) 'Google Earth Engine: planetary-scale geospatial analysis for everyone', *Remote Sensing of Environment*, 202: 18–27.

Hare, F.K. (1993) *Climate and Desertification: A Revised Analysis*, Geneva: World Climate Applications Programme, World Meteorological Organization.

Hurni, H. et al (eds) (2016) *Soil and Water Conservation in Ethiopia: Guidelines for Development Agents*. Bern: Center for Development and Environment.

Kindu, M. et al (2015) 'Drivers of land use/land cover changes in Munessa-Shashemene landscape of the south-central highlands of Ethiopia', *Environmental Monitoring and Assessment*, 187: 452.

Kuma, H.G., Feyessa, F.F. and Demissie, T.A. (2022) 'Land-use/land-cover changes and implications in Southern Ethiopia: evidence from remote sensing and informants', *Heliyon*, 8(3): e09071.

Muñoz Sabater, J. (2019) 'ERA5: fifth generation of ECMWF atmospheric reanalyses of the global climate. Copernicus Climate Change Service Climate Data Store (CDS)'. Available from: https://www.ecmwf.int/en/forecasts/dataset/ecmwf-reanalysis-v5

OECD (Organisation for Economic Co-operation and Development) (2009) *Integrating Climate Change Adaptation into Development Co-operation: Policy Guidance*. Paris: OECD.

Orr, A. et al (2020) 'Smallholder commercialization and climate change: a simulation game for teff in South Wollo, Ethiopia', *International Journal of Agricultural Sustainability*, 19(5–6): 595–608.

Pandey, D. (2008) 'Trees outside the forest (TOF) resources in India', *International Forestry Review*, 10(2): 125–133.

Pellikka, P.K.E. et al (2013) 'Agricultural expansion and its consequences in the Taita Hills, Kenya', in P. Paron, D. Olago and C. Omuto (eds) *Developments in Earth Surface Processes*. Kenya: A Natural Outlook, pp 165–179.

Rahmato, D., Mohammed, Z., Webster, N., Sahle, K., Gizachew, A. and Alemayehu, A. (2021) *Governing Climate Mobility: A Scoping Study of Two Localities in Ethiopia*. DIIS Working Paper 2021: 13. Copenhagen: Danish Institute for International Studies.

SATPALDA (2018) 'Significance of land use/land cover (LULC) maps'. Available from: https://www.satpalda.com/blogs/significance-of-land-use-land-cover-lulc-maps

Sen, P.K. (1968) 'Estimates of the regression coefficient based on Kendall's tau', *Journal of the American Statistical Association*, 63(384): 1379–1389.

Skole, D.L., Samek, J.H., Dieng, M. and Mbow, C. (2021) 'The contribution of trees outside of forests to landscape carbon and climate change mitigation in West Africa', *Forests*, 12(12): 1652.

Thomas, N. et al (2021) 'Trees outside forests are an underestimated resource in a country with low forest cover', *Scientific Reports*, 11: 7919.

Woldie, B.A. and Tadesse, S.A. (2019) 'Views and attitudes of local people towards community versus state forest governance in Tehulederi district, South Wollo, Ethiopia', *Ecological Processes*, 8: 4.

Worku, T., Tripathi, S.K. and Khare, D. (2017) 'Household level tree planting and its implication for environmental conservation in the Beressa Watershed of Ethiopia', *Environmental Systems Research*, 6: 10.

Yinebeb, M., Lulekal, E. and Bekele, T. (2022) 'Composition of homegarden plants and cultural use in an indigenous community in Northwest Ethiopia', *Journal of Ethnobiology and Ethnomedicine*, 18: 47.

PART II

Local Impacts and Adaptation Strategies

4

Assessing Adaptive Capacity to Climate Change and Variability in the Savannah and Forest Agro-Ecological Zones of Ghana

*Francis Xavier Jarawura, Joseph Kofi Teye,
Lily Salloum Lindegaard and Nauja Kleist*

Introduction

Ghana has been considered one of the climate change 'hotspots' in West Africa (Antwi-Agyei et al, 2014). The discourse and research on climate change impacts and adaptive capacities have been concentrated on the north of the country, but growing evidence shows unprecedented and perhaps unexpected challenges in the southern part as well (Teye et al, 2021). While differential physical impacts between the Upper West region and the Eastern region provide insights into the changing nature of spatial vulnerability to climate change in Ghana, they also draw attention to the social dimensions of vulnerability that indeed mediate the outcomes of the physical threats, stresses and shocks to livelihoods (Adger, 2003; IPCC, 2014). A key distinguishing factor between the two regions is that the north has persistently experienced higher poverty and lower levels of development than the south, which means that it currently faces and will likely continue to face far more detrimental effects as capacities to adapt are lower (Teye et al, 2021). Studies detailing such differences in adaptive capacities and the factors influencing them across different climatic zones are rare.

This chapter therefore explores differences as well as similarities in adaptive capacities among households in these two agro-ecological zones: the Savannah (Upper West region) and the Forest zones (Eastern region). Adaptive capacity refers to the potential, capability or ability of any units to adjust to climate change stimuli or to their effects or impacts (IPCC, 2001). Adaptive capacities rest on the nature of household and community assets and broader policies of adaptation (Ellis, 2000; Barnes et al, 2020). Some studies on adaptive capacity have received criticism for a narrow focus on the unit of analysis – for example, individual or household – and failure to consider structural and cross-scalar factors. In line with our attention to governance, we take a broader approach: we take a cross-scalar approach examining adaptive capacity as the interaction between households' own preferences and strategies, and the institutions and governance context that shape the options available to them. This is based on an understanding of vulnerability and capacity to act that hinges not only on wider societal factors, such as the physical properties of the environment, but also on the socioeconomic context and the social preferences of actors (Nielsen and Reenberg, 2010; Rao et al, 2019).

We operationalize this approach through examining the adoption of adaptation practices and the role of institutions in shaping adaptation options and context, which will be detailed in the following section. This approach supports greater scholarly examination of the relationship between climate change, mobility and governance. Specifically, its use of adaptive capacity and adoption literature builds on recent literature linking adaptive capacity and climate mobility (Zickgraf, 2021; Koubi et al, 2022). Our approach further explores this link and illustrates how such an analysis can be carried out. In particular, it illustrates how an adaptive capacity approach to climate mobility: (i) provides insight into the context in which climate mobility decisions are taken, including institutional aspects and differentiated household strategies; and (ii) integrates mobility as an existing aspect of adaptive capacity (see Koubi et al, 2022).

Adaptive capacity: adoption and institutions

The impacts and ramifications of climate change will depend on the ability of human societies to appropriately respond (IPCC, 2001, 2014), contingent upon the effectiveness of societies' adaptive capacity. The IPCC (2001: 982) defines adaptive capacity as 'the ability of a system to adjust to climate change (including climate variability and extremes), to moderate potential damages, to take advantage of opportunities, or to cope with the consequences'. Adaptive capacity encapsulates the ability to make critical and long-term changes to livelihood portfolios. Long-term changes to livelihoods, also referred to as livelihood adaptation, is an enduring 'process of changes to livelihoods intended either at improving existing security and wealth or reducing vulnerability and poverty' (Davies and Hossain, 1997: 5).

Determinants of adaptation and the specific forms they assume vary across space and time. We argue that these are produced through the interplay between a range of factors at and across different scales. For instance, the asset base of households, which is critical to adaptive capacity, is mediated by both internal socioeconomic factors and external trends and shocks (Ellis, 2000). Assets (including human, financial, physical and social assets) are critical in terms of the choices people make of different livelihood strategies when faced by climate and other pressures. Tenure rights, household size, farming experience, wealth, water and extension, access to credit, and off-farm activities are key factors that determine households' ability to adopt and maintain livelihood strategies (Bawakyillenuo et al, 2016). Many of these are influenced by a range of governance and institutional factors, from land and resource management to the provision of extension services. It is also argued that gender is essential in determining access levels to different resources, particularly in rural communities (Partey et al, 2020). This easily finds expression in patriarchal societies such as northern Ghana, where males control production resources (Jarawura and Smith, 2015). Antwi-Agyei et al (2015) also argue that agro-climatic conditions, information and skills, and availability of water are important determinants of adaptive capacity.

These examples of the cross-scalar factors influencing adaptation align with the determinants of adaptive capacity long underscored by the IPCC (2001): economic resources, technology, information and skills, infrastructure, institutions and equity. To operationalize our cross-scalar approach to adaptive capacity, we employ literature on the adoption of adaptation strategies, and institutional and governance factors' influence on these. Through this dual approach, we can examine the interaction between household and community adoption of particular livelihood and adaptation strategies as well as the role of institutions in shaping the adaptation context and options.

Regarding adoption, we consider the key determinants of households', individuals' and communities' adoption of key climate adaptation strategies in the Upper West region and the Eastern region of Ghana. We examine major factors that shape household and community adoption of irrigation, fertilizer use, agronomic practices, nonfarm strategies and migration. Community, household and individual level differences are key to determining differential adoption of various climate adaptation strategies (Epule et al, 2023). These differences are linked to social differentiation and inequalities, which are crucial in creating asymmetrical patterns of adaptive capacities of various actors (see, for example, Adger, 2003; Ribot, 2017; in relation to mobility, see Zickgraf, 2021). Thus, we seek to explain variations in levels of adaptive capacity using a range of individual or household-specific variables, including those of age and gender. In addition, we consider the role of existing mobility practices as part of adaptive capacity, as the two case regions are linked through largely north-south migration streams where there is much interdependence, including sharing of wealth, knowledge and skill. We also examine the role of institutions, which are critical in setting the norms and plans for community, household and individual climate actions. Institutions can be understood as the 'norms, values and practices that guide formal and informal organizations' (Eriksen et al, 2015: 527). They regulate behaviour patterns and also mediate access to key resources such as land, information, markets, infrastructure and technology that are needed for intensification or diversification of livelihood activities (Ellis, 2000, Eriksen et al, 2015).

Our cross-scalar approach to adaptive capacity in the context of climate change and mobility also seeks to provide nuanced insights into why some migrate while others do not. This has been a weakness in some existing literature, in which overly structural accounts have failed to explain differentiated mobility responses to climate change (Koubi et al, 2022), while individual-level analyses are not well suited to account for the institutional or structural factors influencing climate-related mobility (Webster, 2023). While some prominent approaches to climate-related mobility have explicitly taken a cross-scalar approach, notably the Foresight model (Foresight, 2011), this is not well supplemented by an understanding of adaptation and adaptive capacity. The analysis we provide here seeks to address this and can provide a contextual analysis for areas experiencing climate-related mobility.

Study areas in the Upper West region

Two districts were studied in the Upper West region. These are the Wa West and Jirapa districts as shown in Figure 2.1 (see Chapter 2). The topography of both districts is generally flat with few highlands, and it is mainly drained by the Black Volta River and its tributaries. The Black Volta and its tributaries provide opportunities and potentials for irrigation. As in the rest of the Savannah, the two districts experience a single rainy season which lasts from April to September, with average annual rainfall of about 1,150 mm. Rainfall is the major source of watering for agriculture. Farming is a major source of livelihood for both districts with a much greater intensity in the Wa West district (86.0 per cent) than the Jirapa district (67.1 per cent). Farmers in both districts grow a variety of crops, including maize, groundnuts, sorghum, rice, bambara beans and soya beans. They also rear livestock including goats, sheep, pigs, cattle and poultry (GSS, 2012). Besides farming, many people in the region also engage in seasonal, circular and permanent forms of migration, mostly to the southern part of the country (van der Geest, 2010).

Study areas in the Eastern region

In the Eastern region, two districts were studied. These are the Fanteakwa and the Yilo-Krobo districts as shown in Figure 2.1. As the name suggests, the Eastern region is located in the eastern part of Ghana. Its topography is diverse with low-lying areas and highlands. The region falls within the Forest agro-ecological zone. This zone is characterized by dense forest and a bimodal rainfall pattern which occurs from March to October, and from November to February. The mean annual rainfall is around 1,800 mm (Ofori-Sarpong and Annor, 2001; GSS, 2014). The Yilo-Krobo and Fanteakwa districts thus enjoy a double maxima rainfall pattern, which makes farming less risky for people compared to the Savannah, as risk is spread between two cropping seasons rather than one. Agriculture is the major source of livelihoods among people of the Fanteakwa and Yilo-Krobo districts. They mainly grow cocoa, plantain, palm nut, banana, pineapple, cocoyam, maize and mango (GSS, 2014).

Methods of research

This chapter draws on quantitative and qualitative data collected between 2021 and 2023 in the Savannah (Upper West region) and the Forest zone (Eastern region). A survey of some 800 households was conducted among 12 villages based on the proportions of their household populations and using a simple random strategy. The survey utilized a semi-structured questionnaire. The major themes covered include economic strategies, perceptions of climate, governance and relations, adaptation to climate change, and migration. Of the respondents, approximately 56 per cent (449) were males, while 43 per cent (339) were females. About half of the respondents (49 per cent) were illiterate, with the remaining generally having a minimal level of education. Also, the majority of (57 per cent) of the households cultivated five or less acres of land. The qualitative data was collected using focus group discussions and individual in-depth interviews. Two focus groups discussions were held in each village: one involving women and the other involving men. Participants consisted of village subchiefs, leaders of associations and ordinary people numbering between 11 and 16 people in total. These discussions were useful for gathering information on general village-level livelihood strategies and the influence of various internal and external factors in both everyday and long-term changes over time. Some of the key issues included changes and shifts in livelihood strategies, governance of local institutions, access to resources, and climate change perceptions and adaptation.

In-depth interviews were conducted in each of the villages to solicit household and institution-level specific information and to understand the intersections of the two units in shaping climate change adaptive capacity. Adaptive capacity is contingent on the capabilities of households which are largely shaped by institutions in many ways such as the governance of resources (Agrawal, 2010). In each village, we interviewed four men and four women household heads. This was essential in gaining diverse perspectives regarding their adaptive capacities, especially in terms of gender relations. We also interviewed two subchiefs in each village to gain information on how local structures and norms mediate the adaptive capacities of different households. Overall, the quantitative and qualitative approach enabled a mix of strategies in terms of understanding how adaptive capacity is shaped among different villages and households in the study areas. While the quantitative data allowed for analysis that revealed the major factors influencing adaptive capacity, the qualitative data provided insights regarding specific processes and interactions leading to different capacities for climate change adaptation.

Logistic regression model

With the aid of binary logistic regression models, we analysed the quantitative data from the survey of 400 farming households. The binary logistic regression examines the relationship between a binary dependent variable and independent variables such as age and sex. Therefore, the binary logistic regression analysis is appropriate for analysing the data because the outcome variable (adoption of an adaptation strategy) has two independent response categories: whether or not farmers adopted a strategy. Thus the variable Ai represents the farmer's choice to adopt a strategy (for example, off-farm strategies) and β Ei represents a vector of explanatory variables that determine the farmer's decision, such as district, age of respondent, educational level, household size, size of land under cultivation, household size, and perceptions of changes in temperature and rainfall. The binary logistic model is as follows:

$$A^*_i = \beta E_i \cdot + \cdot \varepsilon_i$$

where the farmer's decision to adopt Ai = 1 if the household adopts a particular strategy and 0 if the household does not. Different categories represent household characteristics under each vector. For example, five age categories were used to analyse the adoption of irrigation: 18–25, 26–35, 36–45, 46–60 and > 60. For each set of variables under each vector, a reference category was chosen for the analysis.

Adoption of strategies

We utilized the logistic regression model to examine the determinants of adaptive capacity of various climate change adaptive strategies in the study areas.

Determinants of adoption of irrigation

Irrigation has long been considered an effective way of dealing with climate change and variability (Ellis, 2000; Frisvold and Deva, 2013). In line with this thinking, many policies in Ghana have sought to promote irrigation as a way of ensuring the survivability and progress of rural dwellers. However, such policies and the strategies employed have achieved limited success (Laube et al, 2012; SADA, 2016; Lindegaard et al, this volume, chapter 7). This general situation is reflected in the study areas in terms of low level of irrigation adoption. Only 188 (23.86 per cent) had adopted irrigation as a coping strategy for climate and variability change during data collection. There were no significant differences between adoptions in the districts in the two regions. This is probably explained by the general lack of government support and low adaptive capacities, since both regions have the potential for irrigation. Results from the binary logistic regression showed that the main determinants for adopting irrigation strategies were gender, household size, tenure of cultivated land, received government assistance due to climate hazards in the last five years and perception of change in temperature. As seen in Table 4.1, compared to male farmers, female farmers were nearly two times less likely

Table 4.1: Determinants of the adoption of irrigation

Variable	Category	Irrigation		
		Odd ratio	Standard error	P-value
District	Wa West (*reference category*)	1.0000		
	Jirapa	0.947778	0.250642	0.839
	Yilo Krobo	1.018072	0.311541	0.953
	Fanteakwa	1.581031	0.5765	0.209
Age (years)	18–25 (*reference category*)	1.0000		
	26–35	1.185011	0.425660	0.637
	36–45	0.882832	0.332772	0.741
	46–60	0.754944	0.285615	0.457
	> 60	0.912703	0.373337	0.823
Gender	Male (*reference category*)	1.0000		
	Female	0.662804	0.130592	0.037**
Educational level	No formal education (*reference category*)	1.0000		
	Primary	1.106091	0.309312	0.718
	JHS/Middle school	1.016426	0.286837	0.954
	Secondary	0.650722	0.270841	0.302
	Tertiary	0.374410	0.234670	0.117
Household size	< = 5 members (*reference category*)	1.0000		
	6–10 members	1.385154	0.263066	0.086*
	>10 members	0.495410	0.495491	0.119

(continued)

Table 4.1: Determinants of the adoption of irrigation (continued)

Variable	Category	Irrigation		
		Odd ratio	Standard error	P-value
Land size	< = 5 acres (*reference category*)	1.0000		
	6–10 acres	1.250521	0.269547	0.300
	>10 acres	1.406728	0.353854	0.175
Origin	Native (*reference category*)	1.0000		
	Migrant	1.128227	0.270002	0.614
Tenure of cultivated land	Owned land (*reference category*)	1.0000		
	Leased land	1.230253	0.435283	0.558
	Sharecropped land	0.263517	0.097833	0.000★★★
	Forest reserved land	0.224864	0.171580	0.051★
	Licensed land	0.279563	0.202361	0.078★
	Family land	0.827865	0.184406	0.396
Received government assistance due to climate hazards in the last five years	No (*reference category*)	1.0000		
	Yes	0.633896	0.144486	0.045★★
Perception of occurrence of rainfall	Increasing (*reference category*)	1.0000		
	Same	1.243098	0.328939	0.411
	Reducing	0.440483	0.556475	0.516
Perception of change in temperature hazard	Increasing (*reference category*)	1.0000		
	Same	2.306523	0.56949	0.001★★
	Reducing	0.597467	0.408375	0.451
Observations	788			
Pv	0.0000			
Pseudo R^2	0.0941			

Note: ★ represents a weaker significance level, that is, 10 per cent, and a p-value of 0.05 or greater, but less than 0.1. ★★ denotes a 5 per cent significance level (a p-value of less than 0.05, but greater than or equal to 0.01). ★★★ denotes a 1 per cent significance level (a p-value of less than 0.01). The Pseudo R^2 shows the extent to which the independent variables explain the dependent variable. Thus, a model with a higher Pseudo R^2 is usually preferred to one with a lower Pseudo R^2.

to adopt irrigation as a strategy (p = 0.037). This is despite many efforts by nongovernmental organizations (NGOs) to help women adopt irrigation in order to reduce their general vulnerability and poverty in the society. 'The NGOs always come to help mostly the women to cultivate vegetables during the dry season but we the men have to struggle on our own as they older men say the vegetables farming is a woman's thing, but we still hide behind the women to also benefit', explains an elderly respondent from the Jirapa district during an interview. Differences in irrigation adoption may also be linked to gendered farming practices mentioned by this respondent, where women typically cultivate plots of vegetables, while men cultivate staple crops.

Compared to household size with five or fewer members, households with six to ten members were more likely to adopt irrigation ($p = 0.086$). Again, compared to owned land users, shared crop land users, forest reserved land and licensed land users were about four times less likely to adopt irrigation: ($p = 0.000$), ($p = 0.051$) and ($p = 0.078$), respectively. Interestingly, farmers who had received government assistance due to climate hazards in the last five years were about half as likely to adopt irrigation as a coping strategy ($p = 0.045$). Finally, farmers who believed the temperature has remained the same over the years were two times more likely to adopt irrigation strategies relative to farmers who believe the temperature is increasing ($p = 0.001$).

Determinants of agronomic practices

Changes in and adoption of new agronomic practices are a central part of farmers' adaptation to climate variability and change (Mason and D'croz-Mason, 2002). They generally refer to a range of activities, including changes in crop varieties, changes in planting times (early or late), mixed cropping, the use of tractors instead of human labour, and the intensive practice of mixed cropping. In this study we assess only the adoption of improved or new seed varieties. This is due to the particular attention given to this strategy in focus group discussions and interviews across the two regions:

> Everything we do on the farm matters only when the right type of seed is used and improved seeds and type of seed is what determines the other practices on the farm including when you sow [planting time] and which type of crops you can mix on the plot. So the type of seed is the most important thing for us, the farmers. (Male focus group, Jirapa district)

About 94 per cent of farmers had implemented agronomic practices at the time of data collection. Despite this high proportion, variations were still present across the various districts. As shown in Table 4.2, the main determinants of adopting agronomic practices were district, educational level, land size, cultivated land tenure, perception of rainfall occurrence and perception of temperature change. Compared to the Wa West district, farmers in the Yilo Krobo district were nearly three times less likely to

Table 4.2: Determinants of the adoption of agronomic practices (adopting improved or new crop varieties)

Variable	Category	Agronomic practices		
		Odd ratio	Standard Error	P-value
District	Wa West (*reference category*)	1.0000		
	Jirapa	0.679623	0.161660	0.104
	Yilo Krobo	0.355801	0.103234	0.000★★★
	Fanteakwa	1.032288	0.382123	0.932
Age (years)	18–25 (*reference category*)	1.0000		
	26–35	1.175513	0.412969	0.645
	36–45	1.089097	0.391645	0.812
	46–60	0.916997	0.324549	0.807
	> 60	0.985722	0.380054	0.970
Gender	Male (*reference category*)	1.0000		
	Female	0.786099	0.134428	0.159
Educational level	No formal education (*reference category*)	1.0000		
	Primary	1.164901	0.311421	0.568
	JHS/Middle school	1.53631	0.389427	0.090★
	Secondary	1.489626	0.547528	0.278
	Tertiary	1.992045	0.989524	0.165
Household size	< = 5 members (*reference category*)	1.0000		
	6–10 members	1.326829	0.239696	0.117
	> 10 members	1.002624	0.345837	0.994
Land size	< = 5 acres (*reference category*)	1.0000		
	6–10 acres	1.792177	0.357288	0.003★★
	> 10 acres	2.80648	0.788967	0.000★★★
Origin	Native (*reference category*)	1.0000		
	Migrant	1.203163	0.274175	0.417
Tenure of cultivated land	Owned land (*reference category*)	1.0000		
	Leased land	1.784067	0.780887	0.186
	Sharecropped land	0.750428	0.230543	0.350
	Forest reserved land	1.273022	0.565394	0.587
	Licensed land	0.307352	0.157253	0.021★★
	Family land	1.00998	0.210263	0.962

Table 4.2: Determinants of the adoption of agronomic practices (adopting improved or new crop varieties) (continued)

Variable	Category	Agronomic practices		
		Odd ratio	Standard Error	P-value
Received government assistance due to climate hazards in the last five years	No (*reference category*)	1.0000		
	Yes	0.888651	0.204619	0.608
Perception of occurrence of rainfall	Increasing (*reference category*)	1.0000		
	Same	2.21238	0.60708	0.004**
	Reducing	7.87926	8.08049	0.044**
Perception of change in temperature	Increasing (*reference category*)	1.0000		
	Same	0.481542	0.113819	0.002**
	Reducing	0.188104	0.094554	0.001**
Observations	788			
Pv	0.0000			
Pseudo R²	0.1077			

Note: *, ** and *** indicate significance at the 1, 5 and 10 per cent levels, respectively.

adopt agronomic practices (p = 0.000). Farmers with Junior High School (JHS) or Middle Level Education (MLE) were much more likely to adopt agronomic strategies than farmers without formal education (p = 0.090).

The results also show that compared to farmers with five or less acres of land, farmers with a land size of six to ten acres were almost two times more likely to adopt agronomic practices (p = 0.003). Further, farmers with ten acres of land or more were three times more likely to adopt new or improved seed varieties (p = 0.000). Surprisingly, licensed land users were three times less likely to adopt agronomic practices than owned land users (p = 0.021). Also, farmers who perceived rainfall occurrence as the same and decreasing over the years were, respectively, two times (p = 0.004) and nearly eight times (p = 0.044) more likely to adopt agronomic practices than farmers who perceived rainfall occurrence to increase. This shows the significance of climate perception in farmers' decisions to adopt certain strategies. 'These days we don't get enough rains and when it falls it is not at the right times always as we used to know it', says an elderly woman in Yilo Krobo, Eastern region. Similar expressions were made in the Upper West region: 'Our rains are reducing but the main problem is the way it rains, it does not follow good order so that we can continue to follow our

Table 4.3: Determinants of the adoption of fertilizer application

Variable	Category	Fertilizer application		
		Odd ratio	Standard Error	P-value
District	Wa West (reference category)	1.0000		
	Jirapa	0.688392	0.176115	0.144
	Yilo Krobo	1.477681	0.511950	0.260
	Fanteakwa	0.886875	0.360451	0.768
Age (years)	18–25 (reference category)	1.0000		
	26–35	1.754219	0.653215	0.131
	36–45	1.209321	0.446260	0.607
	46–60	1.472227	0.546754	0.298
	> 60	1.009961	0.407854	0.980
Gender	Male (reference category)	1.0000		
	Female	0.952360	0.188726	0.805
Educational level	No formal education (reference category)	1.0000		
	Primary	1.44807	0.443240	0.226
	JHS/Middle school	1.597277	0.472293	0.113
	Secondary	2.382382	1.102854	0.061
	Tertiary	4.082101	2.491766	0.021**
Household size	< = 5 members (reference category)	1.0000		
	6–10 members	1.325618	0.275449	0.175
	>10 members	1.130271	0.463007	0.765
Land size	< = 5 acres (reference category)	1.0000		
	6–10 acres	1.553766	0.366910	0.062*
	> 10 acres	0.969736	0.271081	0.912
Origin	Native (reference category)	1.0000		
	Migrant	0.513407	0.128706	0.008**
Tenure of cultivated land	Owned land (reference category)	1.0000		
	Leased land	3.609469	1.944894	0.017**
	Sharecropped land	1.446514	0.562032	0.342
	Forest reserved land	0.815837	0.460590	0.718
	Licensed land	0.740485	0.372284	0.550
	Family land	0.812291	0.193965	0.384

Table 4.3: Determinants of the adoption of fertilizer application (continued)

Variable	Category	Fertilizer application		
		Odd ratio	Standard Error	P-value
Received government assistance due to climate hazards in the last five years	No (*reference category*)	1.0000		
	Yes	0.525818	0.152554	0.027**
Perception of occurrence of rainfall	Increasing (*reference category*)	1.0000		
	Same	0.565871	0.160317	0.044**
	Reducing	1.391058	1.027955	0.655
Perception of change in temperature hazard	Increasing (*reference category*)	1.0000		
	Same	1.039564	0.279405	0.885
	Reducing	0.341422	0.166436	0.027**
Observations	788			
Pv	0.0000			
Pseudo R²	0.0873			

Note: * and ** indicate significance at the 1 and 5 per cent levels, respectively.

order of farming as well' (male participant, FGD, Wa West, Upper West region). Moreover, compared to farmers who perceived temperature change as increasing, farmers who perceived change in temperature as the same and decreasing were respectively two times (p = 0.002) and five times (p = 0.001) less likely to adopt agronomic practices.

Determinants of adoption of fertilizer use

The application of fertilizers on farms in Africa has increased significantly over the last century due largely to soil impoverishment (van Heerwaarden, 2022). Respondents point to soil infertility as the main reason for the adoption of fertilizers. 'The soils have become so impoverished as we use it every year. So these days if you don't have money to buy fertilizer, you don't dare farm because you will have a very poor harvest' explains a farmer from the Wa West district in the Upper West region. Farmers in the Eastern region also recognize a decline in soil fertility which they claim has contributed to agricultural extensification.

As shown in Table 4.3, the main determinants of fertilizer use were educational level, land size, origin, tenure of cultivated land, government assistance due to climate hazards in the last five years, perception of rainfall occurrence and perception of change in temperature hazard. The results revealed that farmers with tertiary education were four times more likely to apply fertilizers on their farms than farmers without formal education (p = 0.021). Additionally, farmers who had over ten acres of land were two times more likely to adopt fertilizer applications than farmers with five acres of land size or less (p = 0.062). Moreover, compared to local farmers, migrant farmers were two times less likely to apply fertilizer to their farms (p = 0.008). This is probably explained by the general notion held by participants that migrant households were generally poorer than natives. This finds relevance in the literature on migration of poor rural dwellers who migrate to the relatively wealthier southern regions in search of land (see van der Geest, 2011). Also, farmers who perceived there to be a reduction in temperatures were three times less likely to adopt fertilizer applications compared to farmers who perceived temperatures to have been increasing (p = 0.027).

Determinants of adoption of nonfarm activities

The adoption of nonfarm activities in Africa has been on the rise over the last century in response to many threats to agricultural livelihoods and the opportunities provided by growing nonfarm sectors, such as industries and services (Ellis, 2000). Rural people in Ghana are increasingly adopting nonfarm activities in response to climate variability and change. This allows them to diversify and thus spread both their risks and income sources (Yaro et al, 2015). During data collection, about 60 per cent of respondents indicated their engagement in nonfarm activities due to climate change variability. From the binary logistic regression in Table 4.4, the main determinants of off-farm activities were age, educational level, land use tenure, cultivated land tenure and perception of change in temperature hazards. Districts in the Upper West region were more likely to diversify their livelihoods to include nonfarm activities compared to those in the Eastern region. This can be explained by differentials in the rainfall regimes: the Eastern region has a double maxima rainfall regime, while the Upper West has only one. This means that all risk is concentrated in one season in the latter. Also, a long dry season coupled with poorly developed irrigation infrastructure drive people in the Upper West region to diversify. However, households headed by people aged 60 or over were three times less likely to shift to nonfarm activities compared to those with heads aged 18–25 (p = 0.006).

Compared to farmers who had no formal education, farmers who had a formal education were almost three times more likely to adopt off-farm activities (p = 0.015). Also, farmers using forest-reserved lands were four times less likely to shift to nonfarm activities (p = 0.006) compared to owned land users. On the other hand, licensed land tenure users were about three

Table 4.4: Determinants of the adoption of nonfarm activities

Variable	Category	Nonfarm activities		
		Odd ratio	Standard Error	P-value
District	Wa West (*reference category*)	1.0000		
	Jirapa	0.744404	0.168754	0.193
	Yilo Krobo	1.26878	0.365457	0.409
	Fanteakwa	0.769729	0.249189	0.419
Age (years)	18–25 (*reference category*)	1.0000		
	26–35	0.754851	0.272041	0.435
	36–45	0.553828	0.199282	0.101
	46–60	0.613176	0.219111	0.171
	> 60	0.347989	0.134157	0.006**
Gender	Male (*reference category*)	1.0000		
	Female	0.873499	0.143731	0.411
Educational level	No formal education (*reference category*)	1.0000		
	Primary	1.83234	0.480858	0.021**
	JHS/Middle school	1.61933	0.399525	0.051*
	Secondary	2.003398	0.716633	0.052*
	Tertiary	2.896173	1.265514	0.015**
Household size	< = 5 members (*reference category*)	1.0000		
	6–10 members	0.941027	0.162824	0.725
	>10 members	1.380328	0.487680	0.362
Land size	< = 5 acres (*reference category*)	1.0000		
	6–10 acres	0.819235	0.155384	0.293
	>10 acres	0.727305	0.165884	0.163
Origin	Native (*reference category*)	1.0000		
	Migrant	0.926698	0.190564	0.711
Tenure of cultivated land	Owned land (*reference category*)	1.0000		
	Leased land	0.676077	0.238918	0.268
	Sharecropped land	0.883175	0.258618	0.671
	Forest reserved land	0.256179	0.126214	0.006**
	Licensed land	2.653466	1.429026	0.070*
	Family land	1.095219	0.230761	0.666

(continued)

Table 4.4: Determinants of the adoption of nonfarm activities (continued)

Variable	Category	Nonfarm activities		
		Odd ratio	Standard Error	P-value
Received government assistance due to climate hazards in the last five years	No (*reference category*)	1.0000		
	Yes	0.359029	0.084329	0.000★★★
Perception of occurrence of rainfall	Increasing (*reference category*)	1.0000		
	Same	0.787296	0.198891	0.344
	Reducing	0.467544	0.266641	0.183
Perception of change in temperature hazard	Increasing (*reference category*)	1.0000		
	Same	1.156591	0.2710662	0.535
	Reducing	0.247338	0.126699	0.006★★
Observations	788			
Pv	0.0000			
Pseudo R^2	0.0884			

Note: ★, ★★ and ★★★ indicate significance at the 1, 5 and 10 per cent levels, respectively.

times more likely to shift to nonfarm activities (p = 0.070) than owned land users, and farmers who had received government assistance due to climate hazards in the last five years were three times less likely to shift to nonfarm activities (p = 0.000) compared to farmers who had not received such government assistance. Finally, farmers who perceived temperature as declining were four times less likely to shift to nonfarm activities (p = 0.006), compared to those farmers who perceived rising temperatures.

Determinants of adoption of migration strategies

Migration is increasingly gaining prominence in the livelihood portfolios of rural people in Africa (McAuliffe and Ruhs, 2017). Growing challenges to the farm sector from climate variability and change has been one of the driving forces behind this trend (Adepoju, 2010). The study revealed that 59.90 per cent of respondents reported the adoption of migration strategies as a way of dealing with climate change. The main migration types included were rural-urban migration and urban-rural migration. Districts, age, educational level, origin, tenure of cultivated land, perception of rainfall occurrence and perception of temperature change were the main determinants of migration. As shown in Table 4.5, respondents in districts in

Table 4.5: Determinants of the adoption of migration strategies

Variable	Category	Migration strategies		
		Odd ratio	Standard Error	P-value
District	Wa West (*reference category*)	1.0000		
	Jirapa	2.189197	0.52897	0.001**
	Yilo Krobo	0.8773946	0.2464426	0.641
	Fanteakwa	0.58973	0.1917482	0.104
Age (years)	18–25 (*reference category*)	1.0000		
	26–35	0.6066807	0.2017838	0.133
	36–45	0.514659	0.1757983	0.052*
	46–60	0.6419127	0.2178915	0.192
	> 60	0.4418796	0.1611754	
Gender	Male (*reference category*)	1.0000		
	Female	0.7966838	0.1342696	0.177
Educational level	No formal education (*reference category*)	1.0000		
	Primary	0.6802104	0.1781288	0.141
	JHS/Middle school	0.9484285	0.2262422	0.824
	Secondary	0.5374006	0.1758095	0.058**
	Tertiary	0.1928254	0.0794653	0.000***
Household size	< = 5 members (*reference category*)	1.0000		
	6–10 members	1.078866	0.1927371	0.671
	>10 members	1.600856	0.5602992	0.179
Land size	< = 5 acres (*reference category*)	1.0000		
	6–10 acres	1.356297	0.2626165	0.116
	>10 acres	1.302872	0.2903633	0.235
Origin	Native (*reference category*)	1.0000		
	Migrant	1.945913	0.4104907	0.002**
Tenure of cultivated land	Owned land (*reference category*)	1.0000		
	Leased land	0.7547114	0.2642813	0.422
	Sharecropped land	0.5284162	0.1547634	0.029**
	Forest reserved land	1.758362	0.8329611	0.233
	Licensed land	0.3732113	0.1844112	0.046**
	Family land	1.166608	0.2443868	0.462

(continued)

Table 4.5: Determinants of the adoption of migration strategies (continued)

Variable	Category	Migration strategies		
		Odd ratio	Standard Error	P-value
Received government assistance due to climate hazards in the last five years	No (*reference category*)	1.0000		
	Yes	0.9551056	0.2031805	0.829
Perception of occurrence of rainfall	Increasing (*reference category*)	1.0000		
	Same	0.4392656	0.1196844	0.003**
	Reducing	0.2991418	0.2180636	0.098
Perception of change in temperature	Increasing (*reference category*)	1.0000		
	Same	2.03024	0.4839767	0.003**
	Reducing	1.060733	0.5407217	0.908
Observations	788			
Pv	0.0000			
Pseudo R²	0.1029			

Note *, ** and *** indicate significance at the 1, 5 and 10 per cent levels, respectively.

the Upper West region were more likely to adopt migration than those in the Eastern region. This can be explained by the former's long period of neglect by the state (Yaro et al, 2015), single-maxima rainfall regime (Songsore and Denkabe, 1988) and high levels of poverty (GSS, 2018). The results from Table 4.5 show that compared to the Wa West district, farmers in the more rural Jirapa district were two times more likely to adopt migration in order to cope with climate change (p = 0.001).

Farmers aged between 36 and 45 were nearly two times less likely to adopt migration as a climate coping strategy (p = 0.052) compared to farmers aged between 18 and 25. Farmers with secondary and tertiary levels of education were respectively about two times (p = 0.058) and five times (p = 0.000) less likely to adopt migration as a means of dealing with the challenges that come from climate change than farmers with no formal education. Migrant farmers were almost two times more likely to adopt migration as a coping strategy for climate and variability (p = 0.002) than native farmers. Farmers with sharecropped land and licensed land were respectively two times (p = 0.029) and three times (p = 0.046) less likely to adopt migration strategies compared to owned land users. Also, farmers who perceived rainfall occurrence as being the same were two times less likely to adopt migration as a coping strategy to challenges of climate change (p = 0.003) compared to farmers who perceived rainfall occurrence as increasing, while farmers who perceived temperature as the same were two times more likely to adopt migration as a strategy to climate change ($p = 0.003$) compared to farmers who perceived temperature change as increasing.

Reflections

Through our analysis, we find that location in an agro-ecological zone and household characteristics determine capacities and decisions to adopt various adaptation strategies to climate change. Residence in particular agro-ecological zones (that is, district location) especially determines capacities to adopt, the agronomic strategy of fertilizer application, and migration. Farmers in the Yilo-Krobo district in the Forest zone were more likely to adopt the application of fertilizer. In terms of migration, households in the Savannah were more likely to adopt migration. The importance of location can be found in both the geophysical characteristics, especially climate, and the general socioeconomic circumstances of the particular zone, which shape the possibility of adopting innovation. As was noted earlier, households in the Forest zone have a unique distinction of experiencing a double-maxima rainfall regime that allows them to cultivate crops twice in a year. This allows them to spread risk and increase income. On the other hand, households in the Savannah zone experience only one rainy season, thereby making them relatively more vulnerable to the vagaries of the weather and climate change.

The relative higher wealth of households in the Forest zone is not only due to the quality of the environment but also to the concentration of capital and socioeconomic development found there, the roots for which start with the colonial government and continued under subsequent postcolonial governments (Songsore and Denkabe, 1995; van der Geest, 2011). Poverty is more widespread and enduring in the Savannah (GSS, 2018). The environmental disadvantages of households in the Savannah coupled with relatively higher levels of poverty and a lack of robust infrastructure and development influence southwards migration in search of jobs. The findings also indicate the need for education and credit facilities to allow farmers to procure improved or new seeds that are better suited to changes in climate.

Household-level factors also shape the adoption of certain strategies across the two zones. Age determines the ability to adopt nonfarm strategies and migration. The results show that households with older heads (> 60) are less likely to adopt nonfarm strategies. Focus group discussions tell of the increasing drive for autonomous income among younger generations, with ramifications for traditional household cohesion

(Lindegaard and Jarawura, 2024). However, as was noted in the previous section, it is not always the case that nonfarm strategies are taken on by individuals themselves, but rather that household-oriented strategies are developed as a way to diversify and ensure survivability and progression. The analysis also shows that age influences the adoption of migration. Households whose heads were between the ages of 18 and 25 were most likely to adopt migration. The migration literature has noted a preference of young and middle-aged people for migration, in contrast with the older generations that dislike it for many reasons, including the loss of labour and control (van der Geest, 2010).

Gender also influences the ability to adopt irrigation. Despite targeted support of NGOs, female-headed households are less likely to adopt irrigation, suggesting that additional support has not yet been able to overcome existing inequalities. Education influences the ability to adopt new or improved varieties, fertilizer, nonfarm activities and migration. The stark presence of education in the ability to adopt these types of strategy is evidence of its importance in climate change adaptation. It is well established that while local knowledge is crucial, formal education enables farmers to improve their existing knowledge or gain new knowledge and skills faster (Derbile et al, 2016). The influential role of education in adoption of strategies shows how important it is for long-term climate policies and strategies to focus on ensuring at least basic education for the younger generations.

The size of a household's land also informs the ability to adopt new or improved seeds. Access to land is crucial for adoption of new or improved seed varieties, as different types of landscape and soils are suited for different crops and varieties. The tenure of cultivated land is also critical to the adoption of irrigation, improved or new seeds, fertilizer application, nonfarm activities and migration. Overall, the importance of access to land in making various decisions about adaptation is highlighted by the analysis. It is argued that access to land is critical to farmers' adaptation decisions (Murken and Gornott, 2022). Poor access to different types of land and insecurity of tenure serve as disincentives to the adoption of long-term and costly strategies (Mitchell and McEvoy, 2019). Teye et al (2021) argue that the institutional architecture which determines access to land in Ghana is highly biased against females and fails to grant them enough access to suitable lands for innovation. Access to irrigated land and lack of capacity to initiate autonomous irrigation is particularly noted by respondents as a key challenge to building adaptive capacity. Poor implementation of irrigation plans has been the norm across the country since 1957 (ISSER, 2018). In addition, the sustainability of existing irrigation schemes has been a serious challenge (Sarpong et al, 2022). Baldwin and Stwalley (2022) argue that poor institutional management in Ghana has been a major reason for poor access to, and use of, irrigation land. It is therefore important that adaptation policies and plans give prominence to strategies that enable secure access to land for different social groups.

The capacity to adopt a strategy also depends on how people perceive climate. We found that perception of temperature is extremely important in relation to the adoption of irrigation, improved or new seeds, fertilizer, nonfarm activities and migration. The perceptions of rainfall changes also influence the capacity to adapt improved or new crop varieties. This confirms the assertion that the strategies people employ to change their livelihood strategies are based on what they think and believe is happening (Weber, 2010; Popoola et al, 2018). Although perceptions play a key role in shaping farmers' responses, it is only now receiving growing attention (Whitmarsh and Capstick, 2018). Given the pivotal role of perceptions, there is the need to provide accurate and up-to-date climate information to help inform farmers' understanding of climate conditions.

Conclusions

This chapter has explored the differentiated adaptive capacities between the Forest and the Savannah zones overall, and has also assessed the specific factors at the household level that commonly shape adaptive capacity in both zones. Unlike many studies on adaptive capacity that only focus on specific units of analysis, this chapter has employed a broader cross-scalar approach that includes attention to the role of institutional and governance factors in shaping adaptive capacity. Such an approach allows for a much broader understanding of spatial differentiation and synergies. The knowledge thereby gained can inform better policy making.

The chapter has demonstrated the central role of access to environmental and other assets and therefore the role of institutions in building adaptive capacity. The role of institutions defines access along various lines, including gender, lineage and origin, and thereby shapes the range and intensity of strategies that a household can adopt. Gender-wise, females are generally relegated to the background, while males control productive resources. Also, migrants are confronted with unfavourable tenure arrangements that render them vulnerable and make it difficult for them to appropriately adapt. Gradual and consensus-based institutional changes at the local level will be crucial in reducing the inequalities that exist in resource distribution.

In addition to access to assets, households require other conditions such as skills, favourable market conditions and infrastructure such as irrigation schemes. This confirms the assertion by Adger et al (2005) and Yaro et al (2015) that the asset base of households only becomes effective with the right policy and developmental coordination from the state level. Development failures at the national to regional levels in Ghana have ensured that the rural dweller has remained largely constrained in exploring strategies that could lift them out of poverty (Yaro et al, 2015). It is contended that the development landscape is laden with poor macroeconomic and microeconomic conditions, which do not provide the required economic conditions for improved wellbeing. Adaptation policies in Ghana will therefore need to create the necessary conditions that enable households and individuals to effectively utilize available assets. There is also the need to enable the effective coordination of both traditional and modern institutions in creating the local conditions needed for building adaptive capacity of different social groups in different agro-ecological zones.

References

Adepoju, A. (2010) 'Introduction: rethinking the dynamics of migration within, from and to Africa', in A. Adepoju (ed.) *International Migration within, to and from Africa in a Globalised World*, Accra: Sub-Saharan Publishers, pp 9–45.

Adger, W.N. (2003) 'Social capital, collective action, and adaptation to climate change', *Economic Geography*, 79(4): 387–404.

Adger, W.N., Arnell, N.W. and Tompkins, E.L. (2005) 'Successful adaptation to climate change across scales', *Global Environmental Change*, 15(2): 77–86.

Agrawal, A. (2010) 'Local institutions and adaptation to climate change', in R. Mears, and A. Norton (eds) *Social Dimensions of Climate Change: Equity and Vulnerability in a Warming World*, Washington, DC: World Bank, pp 173–178.

Antwi-Agyei, P., Fraser, E., Dougill, A., Stringer, L. and Simelton, E. (2012) 'Mapping the vulnerability of crop production to drought in Ghana using rainfall, yield and socioeconomic data', *Applied Geography*, 32(2): 324–334.

Antwi-Agyei, P., Stringer, L.C. and Dougill, A.J. (2014) 'Livelihood adaptations to climate variability: insights from farming households in Ghana', *Regional Environmental Change*, 14(4): 1615–1626.

Antwi-Agyei, P., Dougill, A.J. and Stringer, L.C. (2015) 'Barriers to climate change adaptation: evidence from northeast Ghana in the context of a systematic literature review', *Climate and Development*, 7(4): 297–309.

Attua, E.M. (2003) 'Land cover change impacts on the abundance and composition of flora in the Densu basin', *West African Journal of Applied Ecology*, 4(1): 27–34.

Baldwin, G.L. and Stwalley, R.M. (2022) 'Opportunities for the scale-up of irrigation systems in Ghana, West Africa', *Sustainability*, 14(14): 8716.

Barnes, M.L. et al (2020) 'Social determinants of adaptive and transformative responses to climate change', *Nature Climate Change*, 10(9): 823–828.

Bawakyillenuo, S., Yaro, J.A. and Teye, J. (2016) 'Exploring the autonomous adaptation strategies to climate change and climate variability in selected villages in the rural northern savannah zone of Ghana', *Local Environment*, 21(3): 361–382.

Davies, S. and Hossain, N. (1997) *Livelihood Adaptation, Public Action and civil Society: A Review of the Literature*, Brighton: Institute of Development Studies.

Derbile, E.K., Jarawura, F.X. and Dombo, M.Y. (2016) 'Climate change, local knowledge and climate change adaptation in Ghana', in J.A. Yaro and J. Hesselberg (eds) *Adaptation to Climate Change and Variability in Rural West Africa*, Cham: Springer, pp 83–102.

Ellis, F. (2000) *Rural Livelihoods and Diversity in Developing Countries*, Oxford; Oxford University Press.

Epule, T.E. et al (2023) 'A new index assessing adaptive capacity across Africa', *Environmental Science & Policy*, 149: 103561.

Eriksen, S.H., Nightingale, A.J. and Eakin, H. (2015) 'Reframing adaptation: the political nature of climate change adaptation', *Global Environmental Change*, 35: 523–533.

Foresight (2011) *Final Project Report: Migration and Global Environmental Change*, London: The Government Office for Science.

Frisvold, G.B. and Deva, S. (2013) 'Climate and choice of irrigation technology: implications for climate adaptation', *Journal of Natural Resources Policy Research*, 5(2–3): 107–127.

GSS (Ghana Statistical Service) (2012) *Population Census Report*, Accra: Ghana Statistical Service.

GSS (2014) *2010 Population and Housing Census, Municipal Analytical Report, Eastern Region*, Accra: Ghana Statistical Service.

GSS (2018) *Ghana Living Standards Survey Round 7 (GLSS7): Main Report*, Accra: Ghana Statistical Service.

IPCC (Intergovernmental Panel on Climate Change) (2001) 'Climate change: impacts, adaptation and vulnerability', in J.J. McCarthy et al (eds) *A Contribution of the Working Group II to the Third Assessment Report of the Intergovernmental Panel on Climate Change*, Cambridge: Cambridge University Press.

IPCC (2014) *Climate Change 2014. Synthesis Report. Summary for Policy Makers*, Cambridge: Cambridge University Press.

ISSER (Institute of Statistical, Social and Economic Research) (2018) *The State of the Ghanaian Economy in 2017*, University of Ghana, Legon: Institute of Statistical, Social and Economic Research.

Jarawura, F.X. and Smith, L. (2015) 'Finding the right path: climate change and migration in Northern Ghana', in F. Hillmann, M. Pahl, B. Rafflenbeul and H. Sterly (eds) *Environmental Change, Adaptation and Migration: Bringing in the Region*, London: Palgrave Macmillan, pp 245–266.

Koubi, V., Schaffer, L., Spilker, G. and Böhmelt, T. (2022) 'Climate events and the role of adaptive capacity for (im)mobility', *Population and Environment*, 43(3): 367–392.

Laube, W., Schraven, B. and Awo, M. (2012) 'Smallholder adaptation to climate change: dynamics and limits in Northern Ghana', *Climatic Change*, 111: 753–774.

Lindegaard, L.S. and Jarawura, F.X. (2024) 'Loss, damage and social cohesion: Impacts and next steps for policy and programming', DIIS Report No. 2024: 09. Copenhagen: Danish Institute for International Studies.

Mason, S.C. and D'croz-Mason, N.E. (2002) 'Agronomic practices influence maize grain quality', *Journal of Crop Production*, 5(1–2): 75–91.

McAuliffe, M. and Ruhs, M. (2017) *World Migration Report 2018*, Geneva: International Organization for Migration.

Mitchell, D. and McEvoy, D. (2019) *Land Tenure and Climate Change Vulnerability*, Nairobi: UN-Habitat.

Murken, L. and Gornott, C. (2022) 'The importance of different land tenure systems for farmers' response to climate change: a systematic review', *Climate Risk Management*, 35: 100419.

Nielsen, J. Ø. and Reenberg, A. (2010) 'Cultural barriers to climate change adaptation: a case study from Northern Burkina Faso', *Global Environmental Change*, 20(1): 142–152.

Ofori-Sarpong, E. and Annor, J. (2001) 'Rainfall over Accra, 1901–90', *Weather*, 56(2): 55–62.

Partey, S.T. et al (2020) 'Gender and climate risk management: evidence of climate information use in Ghana', *Climatic Change*, 158: 61–75.

Popoola, O.O., Monde, N. and Yusuf, S.F.G. (2018) 'Perceptions of climate change impacts and adaptation measures used by crop smallholder farmers in Amathole district municipality, Eastern Cape province, South Africa', *GeoJournal*, 83(6): 1205–1221.

Rao, N., Mishra, A., Prakash, A., Singh, C., Qaisrani, A., Poonacha, P., Vincent, K. and Bedelian, C. (2019) 'A qualitative comparative analysis of women's agency and adaptive capacity in climate change hotspots in Asia and Africa', *Nature Climate Change*, 9(12): 964–971.

Ribot, J. (2017) 'Cause and response: vulnerability and climate in the Anthropocene', in R. Isakson (ed.) *New Directions in Agrarian Political Economy*, Abingdon: Routledge, pp 27–66.

Sarpong, D.B., Mabhaudhi, T., Minh, T. and Cofie, O. (2022) 'Sustainable financing ecosystem for cocoa irrigation in Ghana: a literature review', Colombo, Sri Lanka: International Water Management Institute (IWMI) and CIGAR Initiative on West and Central African Food Systems Transformation. Available from: https://hdl.handle.net/10568/128228

Savannah Accelerated Development Authority (SADA) (2016) *Resources and Master Plan for the Transformation of Agriculture in the SADA Zone*, Tamale, Ghana: SADA.

Songsore, J. and Denkabe, A. (1995) *Challenging Rural Poverty in Northern Ghana: The Case of the Upper-West Region*, Trondheim: University of Trondheim.

Teye, J.K. et al (2021) *Climate Mobility: Scoping Study of Two Localities in Ghana*, DIIS Working Paper, No. 2021: 10, Copenhagen: Danish Institute for International Studies.

Van der Geest, K. (2010) 'Local perceptions of migration from north-west Ghana', *Africa*, 80(4): 595–619.

Van der Geest, K. (2011) 'North-south migration in Ghana: what role for the environment?', *International Migration*, 49(1): 69–94.

Van Heerwaarden, J. (2022) 'The theoretical potential for tailored fertilizer application: the case of maize in Sub-Saharan Africa', *Field Crops Research*, 288: 108677.

Weber, E.U. (2010) 'What shapes perceptions of climate change?', *Wiley Interdisciplinary Reviews: Climate Change*, 1(3): 332–342.

Webster, N. (2023) 'Shaping spaces: governance and climate-related mobility in Ethiopia', *Climate and Development*: 1–11.

Whitmarsh, L. and Capstick, S. (2018) 'Perceptions of climate change', in S. Clayton and C. Manning (eds) *Psychology and Climate Change: Human Perceptions, Impacts, and Responses*, Cambridge, MA: Elsevier Academic Press, pp 13–33.

Yaro, J.A., Teye, J. and Bawakyillenuo, S. (2015) 'Local institutions and adaptive capacity to climate change/variability in the northern savannah of Ghana', *Climate and Development*, 7(3): 235–245.

Zickgraf, C. (2021) 'Climate change, slow onset events and human mobility: reviewing the evidence', *Current Opinion in Environmental Sustainability*, 50: 21–30.

5

Climate Change and Its Implications for Smallholders' Crop Production: Change from Maize to Teff to Haricot Beans in the Shashemene District, Ethiopia

Busha Teshome and Zerihun Mohammed

Introduction

The evidence for climate change is real, and its consequences are being felt globally (Omerkhil et al, 2020). For instance, the increased concentration of greenhouse gases has raised average temperatures and altered the amount and distribution of rainfall globally (Jamieson et al, 2012). Sub-Saharan Africa (SSA) is among the most vulnerable of the world's regions to climate change impacts (Turyasingura et al, 2022), leading to shorter rainfall durations, the greater incidence of disasters such as droughts and floods, and high temperatures resulting in increased evaporation and reductions in soil water moisture (Connolly-Boutin and Smit, 2016). Climate change has especially affected smallholders in SSA, who depend heavily on rain-fed agriculture for their livelihoods (Connolly-Boutin and Smit, 2016). Ethiopia is one of those countries in SSA that are becoming increasingly vulnerable to climate-induced risks (Asfaw et al, 2021; Giovetti, 2022). Studies show that the effects of climate change in general have had an adverse impact on agriculture (Bravo-Ureta et al, 2020). A review of the literature shows that crop yields respond negatively to increases in temperature (Lobell and Field, 2007; Bravo-Ureta et al, 2020).

In Ethiopia, climate change is causing severe risks for poor farmers. The challenge of meeting the ever-increasing demand for food for the growing population will be further exacerbated by environmental hazards (Conway and Schipper, 2011). Agriculture remains the main activity in the Ethiopian economy and contributes on average 44 per cent of gross domestic product (GDP) and supports the 80 per cent of the population that rely on rain-fed agriculture (MoF, 2020). Crop production is the dominant subsector, accounting for more than 60 per cent of agricultural GDP, followed by livestock with 20 per cent. It is estimated that 16.5 million hectares (14.8 per cent) of the country's land area is potentially suitable for agricultural production (Solomon et al, 2021). Smallholder households produce more than 90 per cent of the agricultural output and cultivate more than 90 per cent of the total cropland. However, due to unpredictable rainfall and changes in temperature, crop production has been severely affected, resulting in a decrease in food security. Hence, the food security and livelihoods of millions of rural farm households are at the mercy of the existing climate at every moment (César and Ekbom, 2013). In this regard, climate-change mitigation and adaptation strategies are crucial (Vijaya Venkata Rama et al, 2012).

Climate change adaptation can be defined as 'the ability to respond to challenges through learning, managing risk and impacts, developing new knowledge and devising effective approaches' (Marshall et al, 2010: 5). Smallholder farming households used several adaptation strategies to resist the various risks posed by climate change (Tesfaye and Seifu, 2016). Households use various on-farm adaptation strategies, including planting different crop varieties, incorporating crop residues into the soil and other water conservation practices, changing planting dates, and irrigation (Kutir et al, 2015; Tesfaye and Seifu, 2016). Climate-change adaptation is a very crucial response strategy to cope with the changing climate (Amamou et al, 2018).

Adaptations to climate change cover a wide range of responses, from praying to changing crop types to migration (Kutir et al, 2015) and non-agricultural employment (Ito and Kurosaki, 2009). Ethiopia has also launched the homegrown economic reform (HGER) from 2021–2030 with the central objectives of sustaining rapid growth, maintaining stable macroeconomic environment by reducing debt vulnerabilities, and creating adequate and sustainable job opportunities (HGER, 2021).

The green economy plan is based on four pillars: (i) improving crop and livestock production practices for higher food security and farmers' incomes while reducing emissions; (ii) protecting and re-establishing forests for their economic and ecosystem services, including their use as carbon stocks; (iii) expanding electricity generation from renewable sources of energy for domestic and regional markets; and (iv) modern and energy-efficient technologies in transport, industry and buildings.

Despite the considerable efforts made, farmers remain vulnerable to the impacts of climate change due to the various constraints they face in making a living. In light of this, there is a limited understanding of how climate change affects crop production and adaptation practices in the Shashemene district. The study was carried out in this region because it is particularly susceptible to climate variability, land degradation and the effects of anthropogenic land-use changes, as well as deforestation. As such, the objectives of this study are to comprehend the effects of climate change on crop production, identify changes in crop production and explore how farmers adapt to these impacts. The findings contribute to informing policy formulation directed at enhancing smallholder adaptation strategies for crop production.

On-farm climate-change adaptation strategies

Climate risks, including extreme climatic phenomena such as droughts, floods, riverbank erosion and heatwaves, have adversely affected crop production and the livelihoods of millions of households in Ethiopia (Gedefaw et al, 2018). To combat climate-related risks and impacts on crop production, farmers are continually introducing adaptation measures that are not always similar to others' livelihood pursuits (Amamou et al, 2018). Thus, the level at which a farm is susceptible to an environmental stimulus is directly related to its ability to cope with the detrimental effects (Danso-Abeam et al, 2021). As mentioned, on-farm adaptation strategies include planting different crop varieties, incorporating crop residues into the soil, water conservation practices, changing planting dates and improving irrigation (Kutir et al, 2015; Tesfaye and Seifu, 2016). In addition, off-farm adaptation strategies encompass a range of risk reduction strategies, including engagement in petty trade, daily labour and migration (Sofoluwe et al, 2011; Danso-Abeam et al, 2021).

Adaptation approaches to crop production can be classified into short-term and long-term adaptations (Olesen and Bindi, 2002). Short-term adaptation strategies include changes in sowing dates, use of fertilizers, tillage and irrigation management (Guo et al, 2010). Long-term adaptation may be achieved by land-use changes and crop breeding (Olesen and Bindi, 2002). Adaptation strategies provide important opportunities to tackle climate change and to sustain crop production (Descheemaeker et al, 2016). Adaptation measures in agriculture depend on the attributes of climate change, farm types, locations, and cost to farmers (Wrigley-Asante et al, 2019). Many potential agricultural adaptation options have been suggested, representing measures or practices that might be adopted to alleviate expected adverse impacts. Although there are well-established concerns about the effects of climate change, there is little quantitative information concerning how serious these effects are for the smallholder farming in the Shashemene district. Studies of the farm-level adaptations that farmers make to cope with the impacts of climate change are also lacking. These seriously limit policy formulation and decision making in terms of adaptation and mitigation strategies.

Adaptation is the strategy through which a region or an individual responds to change through either autonomous or planned activities (Descheemaeker et al, 2016). Sometimes it is difficult to distinguish between coping and adaptive measures, and sometimes coping practices may turn into adaptation practices over time (Brown et al, 2018). Four approaches are considered here: (i) reduced vulnerability; (ii) increased resilience; (iii) enhanced adaptive capacity; and (iv) enhanced transformability (Brown et al, 2018). A community's adaptation strategy is determined by the socioeconomic understanding of the wider society regarding systems and access to resources in the community (Grimm et al, 2013).

Thus, the focus of this particular study is to understand local-level adaptation practices, which are more dependent on experience, knowledge and society. Smallholders are dependent on crop farming; however, due to unpredictable rainfall and changes in temperature, crop production has been severely affected, resulting in a decrease in food security in the Shashemene district. Through this study, we aim to elicit the perceptions of the smallholders themselves to changes in crop production and adaptation to climate change in the study area.

Methodology

The study was conducted in the Shashemene district, West Arsi, Oromia regional state of Ethiopia. The Shashemene district is one of ten districts in the West Arsi zone. It is located 250 km southeast of Addis Ababa. Its average elevation is 1,800 metres above sea level and it is located at 5°12'-8°24' N and 34°22'-38°36'E. The area has an annual average precipitation of 1,057 mm, and the mean maximum and minimum temperatures are 10°C and 24°C respectively (Sahle et al, 2022). The study used a mixed-methods research design. Both qualitative and quantitative methods were employed for data collection. Secondary data were collected from published and unpublished reports and project documents, plans and progress reports. Different documents of relevance to the study were also reviewed. The information and data extracted from the documents were integrated and used in the different sections of the report as appropriate. Primary data were collected by applying different tools, including key informant interviews, focus-group discussions, a household survey and observation. Regarding quantitative methods, a semi-structured household survey was used to support the qualitative analyses.

During the qualitative data collection, 38 key informants were interviewed. These informants included local authorities, community leaders, government agricultural extension officials, nongovernmental organizations (NGOs) and community leaders. The aim was to obtain a comprehensive understanding of the impact of climate change and adaptation issues at the district level.

Additionally, 11 focus group discussions were held, with each group comprising six to eight participants, varying based on the village, district and specific topic of discussion. A structured set of questions was utilized to guide all focus-group discussions, focusing on gaining deeper insights into crop production and the adaptive strategies the farmers had adopted. The household survey was conducted with 410 households from the three villages, namely Fiji Sole, Haleche Harabate and One Chef Umbure village, from the Shashemene district. Within each selected village, farmers were randomly chosen using the farmer lists from the agricultural extension officers in each village. Findings from observation of the study area also helped to validate some of the data from other sources. We also collected secondary data on climate (rainfall and temperature data) from the meteorological station. For this analysis, the qualitative data were analysed using thematic sorting. First, the data were transcribed, which involved typing up the fieldnotes with supplementary information from the recordings. The quantitative data obtained from the household survey were coded and analysed using the Statistical Package for Social Science (SPSS) version 20.0. Descriptive statistics were used to calculate frequencies, percentages, means and medians.

Characteristics of smallholder farmers

A total of 410 smallholder farmers were included in the survey, of whom 222 (54.2 per cent) were men and 188 (45.8 per cent) were women. The average age of the sample population was 36, with a range between 18 and 75 years old. The largest proportion of the household heads (64.9 per cent) lay within the productive age (26–45), 21.8 per cent were in the youth range of 18–25, 11.5 per cent in the age range of 46–60 and the remaining 1.8 per cent were over 60. The sample households have a mean of six family members and a mean of 1.76 hectares of land. Regarding land ownership, the study reveals that the majority of respondents (84.6 per cent) stated that they own the land themselves, while 9.8 per cent had received it from their family, and the remaining acquired land through leasing and sharecropping arrangements. A significant proportion of the respondents (73.6 per cent) were native to the area, indicating that the majority of them are living in their birthplaces. The household asset stock determines the household's level and strategy of adaptation to climate change – for instance, having electricity, a savings bank account and access to drinking water. Households with better access to multiple resources and diverse livelihood portfolios are more likely to be able to cope better with climate risks (Brown et al, 2018; Thornton et al, 2018).

A household's access to electricity (52.6 per cent), drinking water (60 per cent), toilet facilities (83.4 per cent) and a bank account (31.8 per cent) was also significant. The main livelihood strategies of the households show

Table 5.1: Demographic characteristics of the respondent households

Variables	N	%
Sex	Male	54.2
	Female	45.8
Mean family size	6	
Mean age of household head	36	
Marital status	18–25	21.8
	26–45	64.9
	46–60	11.5
	> 60	1.8

Source: GCM Household Survey 2021

that 93.3 per cent are engaged in crop production; maize, teff, sorghum and haricot beans are the main crop varieties produced in the area. Climate change is seen to have affected livelihoods of the households significantly over the last ten years.

Table 5.1 presents demographic data related to a sample population, focusing on sex distribution, family size, age of household heads, and marital status.

Smallholder farmers' perceptions of climate change

The survey results showed that farmers had experienced changes to the local climate over the last decade. The findings revealed that respondents perceived an increased number of droughts (91 per cent), greater irregularity and insufficiency of rain (90.1 per cent), and significant temperature changes (93.9 per cent). In addition, other indicators of change were given: declining soil fertility, problems with input purchases, illness, and diseases among livestock (see Table 5.2). Similarly, in most of SSA, increases in the surface air temperature between 1901 and 2005 amounts to 0.9°C in annual mean temperatures over the entire continent (Alcamo et al, 2007). The increase in rainfall intensity has led to incidents of severe summer flooding in SSA (Christensen and Christensen, 2003). Moreover, during in-depth interviews with key informants and focus group discussions, it was repeatedly stated that the temperature has increased compared to 30 years ago. Similar climate change indicators are indicated in the literature (Reidsma et al, 2009; Juana et al, 2013; Descheemaeker et al, 2016; Mekonnen and Kassa, 2019). In SSA, climate change is manifested in various forms, such as short rainfall durations, the incidence of disasters such as droughts and floods, and high temperatures leading to increased evaporation and reductions in

Table 5.2: Perceptions of the effects of climate change by smallholder farmers (%)

Variables	Increasing	Same	Decreasing	Don't know	Total
Drought	91.0	7.6	0.3	1.2	100
Floods	59.7	20.1	19.5	0.6	100
Rains irregular, insufficient or too heavy	90.1	8.6	0.3	1.1	100
Extreme temperatures (high and/or low)	93.9	5.1	0.3	0.8	100
Crop pests/diseases (for example, locusts, rodents, birds and fungi)	83.5	14.8	0.9	0.9	100
Decreasing soil fertility	84.7	13.6	0.9	0.9	100
Problems with input purchase (for example, seeds, fertilizers and raw materials)	93.6	5.6	0.0	0.8	100
Problems with output sales (for example, agricultural, artisan, fish and dairy products)	85.9	12.5	1.5	0.0	100
Livestock diseases	56.4	32.4	10.4	0.8	100
Lack of drinking water	92.9	4.5	2.3	0.3	100
Illness affecting the household/family	64.6	24.7	9.1	1.5	100
High food prices	97.5	2.5	0.0	0.0	100
Lack of employment	97.7	2.3	0.0	0.0	100
Problems with access to farming land	91.3	8.4	0.3	0.0	100
Problems with access to forestry and non-timber forest products (NTFP)	85.2	13.5	1.3	0.0	100
Conflicts over land (land grabbing)	79.8	17.3	2.2	0.7	100
Conflicts over service provision (for example, schools, health clinics, drinking water and electricity)	73.0	26.2	0.0	0.8	100
Lack of government support to help with households' problems	81.5	17.0	1.2	0.3	100

Source: GCM Household Survey 2021

soil moisture (Connolly-Boutin and Smit, 2016). We found that almost all the smallholder farmers we surveyed could point to changes in their local climate over the last decade, regardless of their farming system or the landscape in which they were located. The most commonly cited changes were rising temperatures, more variable rainfall, and changes in the onset and length of the rainy season.

The trend analysis of temperature and rainfall in the Shashemene district demonstrates a consistent increase in temperature and a decrease in rainfall. This analysis is based on observed data from 1983 to 2020. Specifically, the detailed 37-year analysis indicates a rising trend in both annual maximum and minimum temperatures. Figure 5.1 shows the trends in temperature change from 1983 to 2020, highlighting a general increase in annual mean temperatures. Moreover, the fitted linear trends indicate that mean annual maximum and annual mean minimum temperatures have been increasing over time. The figure shows the trend of the annual maximum temperature for the Shashemene district. This has an upward trend with a slope of 0.05, implying that, since 1983, the maximum temperature has increased by 0.05°C every year in this district. The year 2009 was the hottest year, with a temperature of about 30°C.

The graph in Figure 5.2 depicts the trend in annual minimum temperatures for the Shashemene district. It displays an upward trend with a slope of 0.03, indicating that since 1983, the minimum temperature has been increasing by 0.03°C per year in this district. The coldest year was 1985, with an average temperature of 10.8°C. The rise in annual temperature observed in the study area is attributed to an increase in the summer months, which compensates for the slight decrease in other seasons, particularly the autumn months. The study results indicate that precipitation in the region varies significantly.

The results of the data analysis revealed that the generally declining trend in annual rainfall had been recorded for the last three to four decades. The Shashemene district received below-average rainfall in some years. The years 1984, 1991, 2003, and 2009 were well characterized by below-average rainfall (Figure 5.3). Years with near-average rainfall should not necessarily be years with normal crop production. In the Shashemene district, mainly at lower elevations, the onset of rain is also an important factor in crop production. Years in which the rain arrives late and/or there are significant dry spells were reported as years with lower crop production. An example of such a case is 2015, when a shortfall in crop production was reported.

Impacts of climate change on crop production

The key informants interviewed noticed a decrease in crop production in the study area. The qualitative data collected revealed that the delayed and/or shorter rainy seasons, long droughts, decreased winter rainfall and

Figure 5.1: Maximum temperature time series for the Shashemene district (1983–2019)

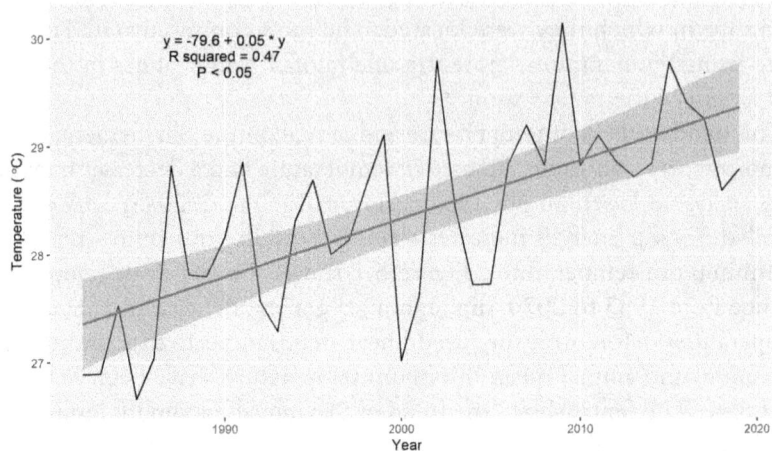

Source: Originally published open access in D. Rahmato, Z. Mohammed and N.A. Webster (2021) 'Governing climate mobility: A scoping study of two localities in Ethiopia', DIIS Working Paper No. 13, Copenhagen: Danish Institute for International Studies

Figure 5.2: Minimum temperature time series for the Shashemene district (1983–2019)

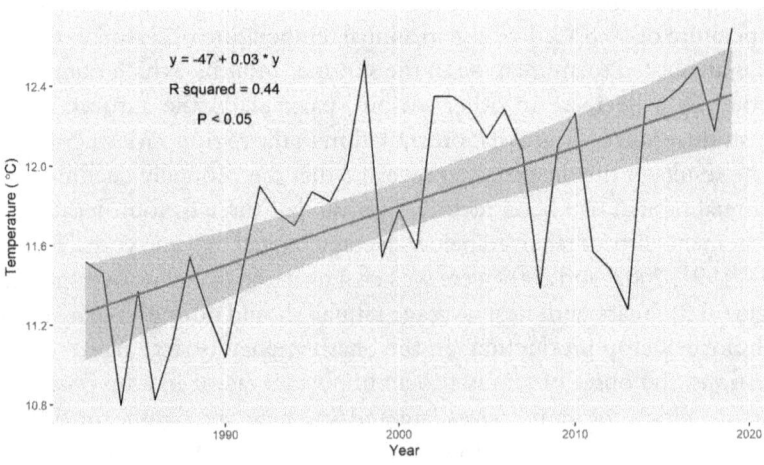

Source: Sahle et al, 2022

increased uncertainty affect crop production in the area. The farmers depend on rain-fed farming, which heavily affects crop production. In addition, the frequency of crop diseases has increased due to the incidence of pests and other climate-related hazards reducing the quality of crops, combined

Figure 5.3: Normalized rainfall anomaly for the Shashemene district (1983–2019)

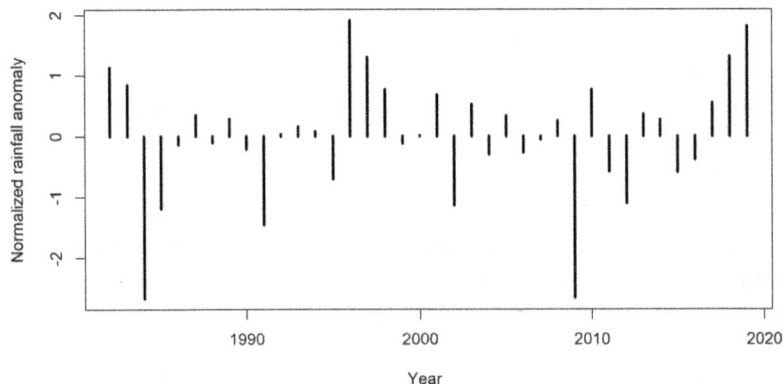

Source: Sahle et al, 2022

with other effects to increase food prices. Decline in incomes has led to a decline in consumption and increased food insecurity. For example, maize production has shown high variability since the 1980s. Maize production decreased from 100 quintals per hectare in 1984 to 50 quintals per hectare in 1994 and 25 quintals per hectare in 2004, before falling even more dramatically to no production in 2021 due to a lack of rain. In addition, crop production was shifted from the long gestation period for crops like maize to short rotation crops such as teff and haricot beans. Over the previous three years, cultivation has shifted to a single crop annually, and even that has failed in some instances. The following statements are illustrative of the farmers' experiences:

> The situation of climate change is getting difficult to survive. Compared to my childhood, currently, there is a shortage and irregularity of rainy seasons, followed by droughts, loss of vegetation, forest and soil erosion, and increasing temperature in the area. After my childhood years (a few years ago), the rain was coming in January; by then, we planted teff and harvested in June. In July, we grew haricot beans again and/or potatoes. The production showed we were producing two times per year, but now even one-time production was impossible. Women like me, particularly those who are household heads, are more vulnerable to climate change than men because I have to feed my seven family members. I am responsible for caring for the children and preparing food for the household, and due to the shortage of water, I am forced to walk a long hour to fetch water for the family. It is an additional burden for me to fetch water from a long distance. Furthermore, in

2020 I planted teff, but due to a shortage of rainfall, it failed to grow. Then I sold one oxen to buy inputs for wheat for planting. Then after planting wheat, I got three quintiles of wheat. The price of wheat was better than the price of maize. I exchanged two quintiles of wheat for two quintiles of maize and 500 birr. In this, I earned 500 Ethiopian birr to buy some items for my family. As a coping mechanism, I engaged in a donkey cart to support my seven family members to buy food, oil, and other things. I prefer Shashemene for working on a donkey cart because it is the nearby town, with availability of daily labouring work and of affordable food to buy for my family in the evening when I go home]. (Yisha Genemo, female household head, at Haleche Harabate, 10 February 2022)

The community in Haleche Harabate has a problem related to climate. The temperature in our village has increased in recent years as compared to 20 years ago, and rain has been short for the last two seasons. For instance, my father had more than 30 livestock and my grandfather had more than 80 livestock and the milk was sold to the nearby town. But now in the village most of the households do not have livestock, and now the community has started to buy milk from the town due to the shortage of grass and water for the livestock to survive in the village. (Abdella Eress, farmer, household head, at Haleche Harabate, 10 February 2022)

Smallholder farmer strategies of adaptation to climate change

To maintain their livelihoods in unfavourable climatic conditions, farmers in the study area have come up with a number of adaptation strategies and practices to cope with the changing climate and to strengthen their resilience (see Table 5.3). The study's findings revealed that farmers' adaptation strategies include adopting an improved or new variety of crop (50.2 per cent), reducing expenses (type and number of meals) (52.4 per cent), using more fertilizer and pesticides (42.2 per cent), using machinery (37.3 per cent), looking for non-agricultural work (27.3 per cent) and seasonal migration (29.3 per cent). Other adaptation strategies in this study include the use of soil-water conservation measures (35.6 per cent) and the use of machinery (37 per cent). Crop and cattle insurance (0.5 per cent) was the least adopted adaptation strategy by farmers. Crop diversification in space (substituting one crop for another) and time (changing the crop rotation or cropping system) can be a rational and cost-effective way to build the resilience of agricultural systems that are subject to climate change (Lin, 2011).

Table 5.3: Climate change adaptation strategies of smallholder farmers

Adaptation strategy for household	% of respondents
Buy crop/cattle insurance	0.5
Improved irrigation	4.9
Short- or long-term migration abroad	9.8
Permanent migration away from settlement by household members (no plan to return in next five years)	10.5
Start to produce handicrafts	15.1
Begin trade in agricultural and other commodities	19.8
Look for non-agricultural work	27.3
Seasonal migration (annually) by household memebers to rural areas	29.3
Maintaining fewer livestock	35.4
Take steps against soil erosion	35.6
Seasonal migration (annually) by household memebers to urban areas	37.1
Farming on less land	37.3
Use more machinery	37.3
Use more fertilizer and pesticides	42.2
Change time of planting	49
Adopting improved or new crop varieties	50.2
Reduce expenses (for example, type and number of meals)	52.4

Source: GCM Household Survey 2021

Lack of insurance and diversification could probably be attributed to the limited technical skills and financial capacity of smallholder farmers to buy adequate insurance. All the adaptation strategies reported focused on reducing the effects of drought, which seem to be a more frequent problem for farmers in the study area. The key informants suggested that when the government, NGOs and the private sector engaged in developing climate-change adaptation strategies, they should therefore consider the context of the community.

Crop production-based climate-change adaptation strategies

Climate change has affected agriculture in African countries by adversely impacting on different farming systems (Gornall et al, 2010). Climate change may require adaptation and changes in farmers' practices to conserve and improve crop yields. The farmers practised short-duration crops, changed the cropping pattern and calendar, and made use of drought-tolerant varieties. Smallholders have tried to use drought-resistant crops as adaptation methods to climate change in the Rift Valley area of Ethiopia (Tofu et al, 2022). Changing cropping patterns, introducing new crops or replacing existing crops, and changing crop sequences are ways of adapting to climate change (Lasco et al, 2011). The key informants confirmed the pattern of change in crop production from long rotation to short rotation for maturity. For instance, maize was planted in May and harvested in October, after six months of growth. However, due to the limited availability of rain, farmers push to plant teff in February and harvest in June. In this period, they were able to plant additional potatoes after harvesting the teff. Recently, even though the rain did not come in February, teff was replaced with short-rotation haricot beans in August, to be harvested in October. When there is rain, they plant wheat in August and harvest in November as well. This trend indicates that the farmers are changing from planting a long-gestation crop like maize or teff to short-rotation crops like haricot beans. The introduction of short-duration crop varieties and the planting of early- and late-maturing varieties may help reduce the adverse impacts of climate risk (Lasco et al, 2011). Replacing this cropping system with less water-intensive cropping patterns (for example, maize to wheat) can enhance the adaptation of the production system to water stress. Similarly, diversifying production systems through the promotion of 'neglected and under-utilized species' offers adaptation opportunities to climate change, particularly in the mountains (Adhikari et al, 2015, 2018).

Timely planting is a farm-management practice that is aimed at getting the crop planted at optimum planting dates largely driven by the prevailing weather conditions (Tambo and Abdoulaye, 2013). The key informant we interviewed confirmed that timely planting was one of the adaptation strategies in the direction of teff and haricot bean production under climate change. The use of drought-tolerant varieties was also confirmed to be an adaptation option, as were haricot beans. Key informants also reported that farmers had started to buy milk from the town, although usually milk was supplied from rural to urban areas. However, due to the drought in rural areas, farmers cannot keep their livestock for milk production.

The possible impacts of climate change on food security have tended to be viewed with most concern in locations where rain-fed agriculture is still the primary source of food and income. The focus group discussions with farmers revealed that they possessed a wide knowledge of adaptation strategies, such as reductions in input requirements and increases in yield production of a crop. The crop comparison is focused on the input-output found in Table 5.4, which provides an indication of the differences in estimates between the two crops. The common approach to calculating

Table 5.4: Comparison of input cost and selling price of maize and teff production (2020)

	Unit	Maize	Teff
Inputs	Fertilizer	250 kg/ha	25 kg/ha
	Seed	24 kg/ha	40 kg/ha
Input cost	US$	183	52
Production	Quintiles	25	16
Price per quintal	US$	57/quintile	96/quintile
Total production	US$	1,425/ha	1,536/ha
Net profit	US$	1,242	1,484
Pest	–	Highly susceptible	Less susceptible
Duration of production	Months	May–October	February–June
Frequency of production	–	Only maize	Allow for other production
Market	–	Available demand	Better market demand

Source: GCM Household Survey and FGDs

the value of each crop applied by the existing studies is to use prices based on key informants' self-reported production and sales information. For instance, farmers have shifted from maize to teff production based on their input costs and price per quintile. The ability to purchase inputs depends on the condition of the household prior to and during the farming season. As indicated in Table 5.4, the price of inputs for a hectare of maize was US$183 per hectare, while the input cost for teff was US$52 per hectare. The selling price was US$57 per quintile for maize and US$96 per quintile for teff, which shows that the price of teff was much more attractive than the price of maize. From the current change in crop production, it is understood that farmers have shifted to producing a crop having low price inputs and with a better sale price. Furthermore, the total income from the production of maize on one hectare is about US$1,425, while for teff it is US$1,536. Therefore, the net benefit from the production of teff is much higher than that for maize and in a shorter period.

Our key informants confirmed that one of their adaptation strategies was to turn to the private sector to work with farmers in crop cultivation. The private sector was turned to in times of drought, which necessitated inputting seed and fertilizer to support the farmers. In the study area, the private company (called SORRETI) was active in supporting crop production as input suppliers and the buyer. The company entered into contractual agreements with the farmers, mostly involving direct agreements to provide farmers with agricultural inputs such as seeds and fertilizer on credit; it then bought the grain at a guaranteed price of its market value at the time of its harvest. This company's arrangement with the farmers' groups included a buy-back agreement, and it also provided them with an advisory service. Accordingly, SORRETI provided 1,200 quintals of haricot beans and fertilizer for 60 clusters with 20 farmers per cluster in 2021. The partnership and production in the cluster is of benefit to farmers in terms of their supporting each other, accessing improved seed and fertilizer, and accessing the market. The support and facilitation of the private sector (SORRETI) is greatly appreciated and acknowledged by the farmers, other private sectors and the government, as they are effective in providing seed and subsequently as a buyer of the crop. Despite being driven by profitability, SORRETI was also addressing the farmers' problems regarding climate change, ensuring the timely provision of seeds and inputs during periods of drought.

Conclusion and policy implications

Most respondent households indicated that they had experienced climate-change effects over the last two decades. Climate change in the area takes the form of increased temperatures, drought, the shortage and variability of rainfall, and new crop diseases in the area. The farmers in the study, who rely substantially on rain-fed agriculture for their sustenance, have been impacted by climate change, leading to a reduction in crop yields. The farmers have adjusted their farming practices to account for the impacts of climate change. The main adaptation strategies of farmers are changing crop varieties, the introduction of short-duration crops, planting low input-cost crops and changing crop-planting dates. The study has significant implications for policy makers, donors and practitioners seeking to enhance the resilience of smallholder farmers in the study area to climate change. It suggests the importance of supporting smallholder farmers through partnerships between the public and private sectors, as well as asking farmers themselves to cope with the current climate challenges and adapt to future climatic conditions. Furthermore, there is a need to consider the farmer's preference for adaptation strategies that are accessible to smallholder farmers and that suit their agro-ecological and socioeconomic contexts. These farmers are using local understandings of adaptation in relation to crop production in the climate-change scenario. Hence, it is also vital for governments to consider smallholder farmers' perceptions of climate change in formulating adaptation policies.

References

Adhikari, U., Nejadhashemi, A.P. and Woznicki, S.A. (2015) 'Climate change and eastern Africa: a review of impact on major crops', *Food and Energy Security*, 4(2): 110–132.

Adhikari, B.N., Joshi, B.P., Shrestha, J. and Bhatta, N.R. (2018) 'Genetic variability, heritability, genetic advance and correlation among yield and yield components of rice (Oryza sativa L.)', *Journal of Agriculture and Natural Resources*, 1(1): 149–160.

Alcamo, J., Flörke, M. and Märker, M. (2007) 'Future long-term changes in global water resources driven by socioeconomic and climatic changes', *Hydrological Sciences Journal*, 52(2): 247–275.

Amamou, H. et al (2018) 'Climate change-related risks and adaptation strategies as perceived in dairy cattle farming systems in Tunisia', *Climate Risk Management*, 20: 38–49.

Asfaw, A., Bantider, A., Simane, B. and Hassen, A. (2021) 'Smallholder farmers' livelihood vulnerability to climate change-induced hazards: agroecology-based comparative analysis in Northcentral Ethiopia (Woleka Sub-basin)' *Heliyon*, 7(4): 1–14.

Bravo-Ureta, B.E., Higgins, D. and Arslan, A. (2020) 'Irrigation infrastructure and farm productivity in the Philippines: a stochastic Meta-Frontier analysis', *World Development*, 135: 105073.

Brown, P.R., Tuan, V.V., Nhan, D.K., Dung, L.C. and Ward, J. (2018) 'Influence of livelihoods on climate change adaptation for smallholder farmers in the Mekong Delta Vietnam', *International Journal of Agricultural Sustainability*, 16(3): 255–271.

César, E. and Ekbom, A. (2013) *Ethiopia Environmental and Climate Change Policy Brief*, Stockholm: Sida's Helpdesk for Environment and Climate Change.

Christensen, J.H. and Christensen, O.B. (2003) 'Severe summertime flooding in Europe', *Nature*, 421(6925): 805–806.

Connolly-Boutin, L. and Smit, B. (2016) 'Climate change, food security, and livelihoods in Sub-Saharan Africa', *Regional Environmental Change*, 16(2): 385–399.

Conway, D. and Schipper, E.L.F. (2011) 'Adaptation to climate change in Africa: challenges and opportunities identified from Ethiopia', *Global Environmental Change*, 21(1): 227–237.

Descheemaeker, K., Oosting, SJ., Homann-Kee Tui, S., Masikati, P., Falconnier, G.N. and Giller, K.E. (2016) 'Climate change adaptation and mitigation in smallholder crop–livestock systems in Sub-Saharan Africa: a call for integrated impact assessments', *Regional Environmental Change*, 16(8): 2331–2343.

Gedefaw, M., Girma, A., Denghua, Y. et al (2018) 'Farmer's perceptions and adaptation strategies to climate change, its determinants and impacts in Ethiopia: evidence from Qwara district', *Journal of Earth Science & Climatic Change*, 9(7): 1000481.

Giovetti, O. (2022) 'Climate change in Ethiopia: what happened in 2021, and what is the forecast for 2022?', *Concern*, 11 February. Available from: https://www.concern.net/news/climate-change-in-ethiopia

Gornall, J. et al (2010) 'Implications of climate change for agricultural productivity in the early twenty-first century', *Philosophical Transactions of the Royal Society B: Biological Sciences*, 365(1554): 2973–2989.

Grimm, N.B. et al (2013) 'The impacts of climate change on ecosystem structure and function', *Frontiers in Ecology and the Environment*, 11(9): 474–482.

Guo, R., Lin, Z., Mo, X. and Yang, C. (2010) 'Responses of crop yield and water use efficiency to climate change in the North China Plain', *Agricultural Water Management*, 97(8): 1185–1194.

HGER (2021) Ten years development plan a pathway to property. Federal Democratic Republic of Ethiopia, Planning and Development Commission.

Ito, T. and Kurosaki, T. (2009) 'Weather risk, wages in kind, and the off-farm labor supply of agricultural households in a developing country', *American Journal of Agricultural Economics*, 91(3): 697–710.

Jamieson, M.A., Trowbridge, A.M., Raffa, K.F. and Lindroth, R.L. (2012) 'Consequences of climate warming and altered precipitation patterns for plant-insect and multitrophic interactions', *Plant Physiology*, 160(4): 1719–1727.

Juana, J.S., Kahaka, Z. and Okurut, F.N. (2013) 'Farmers' perceptions and adaptations to climate change in Sub-Saharan Africa: a synthesis of empirical studies and implications for public policy in African agriculture', *Journal of Agricultural Science*, 5(4): 121–135.

Kutir, C., Baatuuwie, N.B., Keita, S. and Sowe, M. (2015) 'Farmers awareness and response to climate change: a case study of the North Bank Region, the Gambia', *Journal of Economics and Sustainable Development*, 6(24): 32–41.

Lasco, R.D. et al (2011) *Climate Change Adaptation for Smallholder Farmers in Southeast Asia*. Manila: World Agroforestry Centre.

Lin, B.B. (2011) 'Resilience in agriculture through crop diversification: adaptive management for environmental change', *BioScience*, 61(3): 183–193.

Lobell, D.B. and Field, C.B. (2007) 'Global scale climate–crop yield relationships and the impacts of recent warming', *Environmental Research Letters*, 2(1).

Marshall, N.A. et al (2010) *A Framework for Social Adaptation to Climate Change: Sustaining Tropical Coastal Communities and Industries*, Gland, Switzerland: IUCN.

Mekonnen, Z. and Kassa, H. (2019) 'Living with climate change: assessment of the adaptive capacities of smallholders in central rift valley, Ethiopia', *American Journal of Climate Change*, 8(2): 205–227.

MoF (Ministry of Finance) (2020) *Ethiopia's National Economy Technical Report*. Addis Ababa: Ministry of Finance.

Olesen, J.E. and Bindi, M. (2002) 'Consequences of climate change for European agricultural productivity, land use and policy', *European Journal of Agronomy*, 16(4): 239–262.

Omerkhil, N. et al (2020) 'Climate change vulnerability and adaptation strategies for smallholder farmers in Yangi Qala district, Takhar, Afghanistan', *Ecological Indicators*, 110: 108863.

Reidsma, P., Ewert, F., Oude Lansink, A. and Leemans, R. (2009) 'Vulnerability and adaptation of European farmers: a multi-level analysis of yield and income responses to climate variability', *Regional Environmental Change*, 9(1): 25–40.

Sahle, K. et al (2022) *Governing Climate Mobility: Project Report of Scoping Study in Ethiopia*. Copenhagen: DIIS. Available from: https://www.diis.dk/en/research/governing-climate-mobility-0

Sofoluwe, N.A., Tijani, A.A. and Baruwa, O.I. (2011) 'Farmers' perception and adaptation to climate change in Osun State, Nigeria', *African Journal of Agricultural Research*, 6(20): 4789–4794.

Solomon, R., Simane, B. and Zaitchik, B.F. (2021) 'The impact of climate change on agriculture production in Ethiopia: application of a dynamic computable general equilibrium model', *American Journal of Climate Change*, 10(1): 32–50.

Tambo, J.A. and Abdoulaye, T. (2013) 'Smallholder farmers' perceptions of and adaptations to climate change in the Nigerian savanna', *Regional Environmental Change*, 13(2): 375–388.

Tesfaye, W. and Seifu, L. (2016) 'Climate change perception and choice of adaptation strategies: Empirical evidence from smallholder farmers in east Ethiopia', *International Journal of Climate Change Strategies and Management*, 8(2): 253–270.

Thornton, P.K. et al (2018) 'Is agricultural adaptation to global change in lower-income countries on track to meet the future food production challenge?', *Global Environmental Change*, 52: 37–48.

Tofu, D.A., Woldeamanuel, T. and Haile, F. (2022) 'Smallholder farmers' vulnerability and adaptation to climate change induced shocks: the case of Northern Ethiopia highlands', *Journal of Agriculture and Food Research*, 8: 1–23.

Turyasingura, B. et al (2022) 'A systematic review of climate change and water resources in Sub-Saharan Africa'. https://doi.org/10.21203/rs.3.rs-2281917/v1

Vijaya Venkata Raman, S., Iniyan, S. and Goic, R. (2012) 'A review of climate change, mitigation and adaptation', *Renewable and Sustainable Energy Reviews*, 16(1): 878–897.

Wrigley-Asante, C., Owusu, K., Egyir, I.S. and Owiyo, T.M. (2019) 'Gender dimensions of climate change adaptation practices: the experiences of smallholder crop farmers in the transition zone of Ghana', *African Geographical Review*, 38(2): 126–139.

6

Khat Cultivation and Climate Change in Tehuledere, South Wollo

Dessalegn Rahmato

Introduction

Ethiopia, one of the most vulnerable African countries to climate change, has been a frequent victim of extreme weather events, such as droughts, excessive temperature, floods and landslides, often accompanied by outbreaks of livestock disease and pest infestation.[1] Such events have often triggered severe food shortages and famines, large-scale losses of farm animals and human displacement (World Bank, 2010a; Dessalegn et al, 2013; EAS, 2015). The highlands of northeast Ethiopia, where Tehuledere, the woreda (district) where the research for this chapter was conducted lies, have suffered far more severe and more frequent extreme weather events than any other part of the country. The most frequent impact on the region has been drought, which has often led to the collapse of agricultural production and large-scale human suffering.

However, while climate shocks and their consequences have left an indelible scar in the social and ecological landscape of the region, they have also made the rural population more resourceful. Rural households here have learnt to make ingenious use of locally available assets to mitigate the harmful impacts of climate adversity, becoming, in the process, active agents of their own survival rather than the silent victims of natural disasters.

This chapter will look at one example of such resourcefulness in a small community of khat growers in a rural setting in northeast Ethiopia. It will show how these growers have managed to make khat serve a dual purpose, namely as a source of household income on the one hand and as a tool to withstand climate shocks on the other. What we have here is the substance of what is known as autonomous adaption, but which we have renamed indigenous adaptation, and which, as we shall see, consists primarily of farmers' everyday agricultural practices, often renewed by local knowledge and experience. Smallholder farming routinely incorporates adaptation measures and should in fact be called adaptation farming.

Rural livelihoods in Tehuledere

Let us take a brief look at the physical resources of the woreda, information for which was obtained from the Central Statistical Agency (CSA) and the Amhara Region Agriculture Bureau.[2] The CSA's population data show that Tehuledere is the most densely settled woreda in Amhara region, with a population density of 387.1 persons per km^2. But the resources available are insufficient to sustain such a population, and, as we shall see later, they are becoming ever smaller due to demographic, environmental, agronomic and other factors. The vast majority of the population depends on farming for its livelihood, but according to the Agriculture Bureau's information, less than half of the woreda's land is under cultivation, the rest consisting of rugged terrain, hills and mountains. More importantly, the woreda's agro-ecological landscape, on which small farmers depend to plan their farming calendar (which is discussed in some detail later on) is undergoing change and will have a damaging impact on agricultural productivity. While food grains such as sorghum, maize, wheat and teff have long been the main crops in the woreda, fruit and vegetables have been grown as supplementary outputs in recent years. Tehuledere, like the South Wollo Zone as a whole, has been a food-deficit area, and food scarcity has been a prime concern of farmers and others for many generations.

The winds of change have been blowing in Tehuledere for some time, but while some of the population has welcomed many aspects of it, others see it as a challenge and as a threat to their livelihoods. Some of the main drivers of change include urbanization, population growth, state-sponsored investment in a range of public services and infrastructure, and efforts at agricultural improvement. Urbanisation is expanding in our woreda, as well as in the South Wollo Zone; indeed, the growth of towns and cities (both new and old) across the country has been noted by many as an important dynamic of the changes that are taking place in Ethiopia as a whole (see Pankhurst and Dom, 2019). The Mayor of Haiq town, the capital of the woreda, who we interviewed in March 2020, was very positive about the growing urbanization in the woreda in the last two to three decades, although he did recognize that there were challenges that have emerged that need to be addressed. Haiq's residents, he informed us, were composed of full-fledged urban residents, but also a sizeable number are farming people who have one leg in the urban environment and the other in the rural environment. The capital, along with other, smaller towns in Tehuledere, has expanded into the countryside, and this has brought urban ways of living to rural dwellers along with manufactured consumer goods. On the other hand, urban expansion has been at the expense of rural communities since it takes away farmland, forest and water resources, and causes population displacements.

Most of Haiq's residents rely on trade for a living, but there are also people engaged in the expanding transport sector, in hotels, restaurants and food processing, which benefit from expanding transportation; there are also people working in construction, craftsmen and women, and those engaged in repair work of all types. The informal sector is also an important source of livelihood for residents. It consists of a large number of young boys making a living as peddlers, hawkers and shoeshiners. A relatively new source of employment for young women in the informal sector is the 'pavement' or 'sidewalk' cafe, which involves providing coffee, tea and street food to customers in temporary stalls put up on street corners and public spaces. This, we were told, has become quite popular with town residents and visitors, and serves as a source of independent income for women and girls. During our interview, the mayor pointed out that Haiq attracts a large number of rural migrants from the surrounding areas, from the South Wollo Zone and from areas as far away as Tigray.[3]

The urban economy relies heavily on the countryside, which provides agricultural products such as food grain, fruit and vegetables, condiments and livestock, and of course khat, in return for which it receives consumer goods, many of which are imports. The woreda does not produce any exportable goods, not being noted for the production of coffee, oilseeds or livestock products; it is a player only in the local economy, and even so only when it is not falling victim to natural disasters such as drought, which is not very often. Thus, the urban economy serves as the main conduit for the entry of global capital into the rural world. To small farmers in Tehuledere, globalization takes the form of cheap manufactured goods, such as household utensils, ready-made clothes, radios, television sets and mobile phones. A recent addition to this list of imports is the three-wheel taxi, which has become ubiquitous in all towns, villages and settlements in the country that are served by roads. Table 6.1 shows some of the consumer goods that

Table 6.1: Farmers' ownership of selected assets by kebele (%)

Kebele	Oxen	TVs	Radios	Mobile phones
Bededo	32	54	29	84
Hitecha	60	22	29	80
Seglen	52	7	15	83

Source: GCM Household Survey 2021

are owned by rural households as revealed by the GCM Household Survey conducted in three of the kebeles in the woreda. It is interesting that over 80 per cent of respondents said they own mobile phones; this is far higher than the ownership of oxen, which are vital to the main means of farming both here and elsewhere in rural Ethiopia.

Urbanization has gone hand in hand with the expansion of basic infrastructure across rural Ethiopia. Considerable investment has been made in major and secondary roads, hydropower projects, communication systems and, in the case of South Wollo, railway infrastructure. Tehuledere, like other predominantly rural woredas, has benefited from access to electric power, though businesses and local officials in Haiq complain of frequent and unpredictable power cuts. Roads have made transport outside one's locality easier: they have led to greater interaction among populations, which were previously confined to their home areas and have improved the chances of accessing jobs for the rural population, especially in the off-season. Until the 'Northern War' broke out in 2020, the construction of the railway line, which passes near the woreda, was a good source of employment for the surrounding population.

A word is appropriate about urbanization and khat cultivation. Our interviews with khat growers in Tehuledere confirmed that urbanization has opened up a large and growing market for khat. The consumption of khat, especially for recreational purposes, is associated with urban ways of life, and townspeople have greater purchasing power than rural dwellers. Hence, urban demand is high and growing, and urban traders have become the main movers of khat from the countryside to the big urban centres. The flow of the khat trade in the country is predominantly from rural to urban areas, with rural-to-rural exchanges making up only a small portion of the total.[4] As we shall see later in this discussion, the main markets for khat in our woreda are in Haiq and the bigger towns in South and North Wollo located close by. It is thus safe to argue that the growth of urbanization has played a significant role in the expansion of khat cultivation not just in Tehuledere, but also in the country as a whole in the last three decades.

All the evidence available shows that the country's population is growing fairly rapidly, and the majority of the population is young.[5] This has major implications for access to resources, asset distribution and, in the context of the agrarian economy, land tenure. There are two major pressures that come to bear on agricultural land: pressure from within and pressure from outside. Internal pressures are those relating to inheritance claims by siblings and other family members on the one hand, and the demands on the kebele to provide social services on the other hand. In almost all rural communities, family holdings are getting smaller as a result of being subdivided within the family; similarly, farmland and pastureland are expropriated by the kebele to construct schools, health posts and other service institutions, with the result that common lands have disappeared (Dessalegn, 2009). The pressure from outside refers to the demands for land for a wide variety of purposes, including urbanization, the needs of the modern sector of the economy (land for manufacturing enterprises, for industrial parks and so on), the construction of infrastructural projects and other investments (Pankhurst and Dom, 2019).

In several of the communities studied by the WIDE programme, landholdings per family measure well below one hectare, with the mean standing at 0.5 hectares (ha) (Pankhurst and Dom, 2019). According to informants at the land office in Tehuledere, there is a severe shortage of land in the woreda, and landlessness is especially high among the young. Data for the average size of holdings in the three kebeles of our study as shown by the GCM Household Survey reveal miniscule household plots: Bededo = 0.45 ha; Hitecha = 0.54 ha; and Seglen = 0.66 ha. This does not mean there are no families with more land, and in fact the larger holdings reported by a small number of families measure between 1.50 and 2.00 hectares. However, what is clear is that the property regime in Tehuledere is dominated by micro-holdings and, thanks to sibling inheritance rights, has become 'dynamic in reverse' – that is, caught up in the cycle of the levelling down of ever-smaller holdings.

The implementation of land registration and certification in the country since the early 2000s was supposed to reduce the fragmentation of land, as well as disputes over land. There have been two rounds of land registration, but while the first round was carried out in most parts of rural Ethiopia, the second round was still unfinished as of 2022. In Tehuledere, only the first round was undertaken, and this, according to officials at the woreda land office, has given rise to numerous disputes over land. At the time of our visit, a crowd of people were waiting their turn at the land office: some of them were there to lodge new cases of dispute, others to find out what has been decided regarding the cases which they had lodged earlier. A wide range of disputes have been brought to the land office, as well as to the woreda courts, as found in earlier studies in Wollo (Dessalegn, 2009). There are disputes having to do with plot boundaries which were not demarcated properly during the first round of certification. In the second round, plot boundaries were more accurately demarcated with the help of Global Positioning System (GPS) technology, and a map of the plot or plots was attached to each issued land certificate. There are also marital disputes, which invariably involve the issue of property rights; disputes arising out of claims and counterclaims to plots by family members or others; disputes having to do with who is or is not entitled to inheritance; and disputes among siblings over how and when family property should be divided. Land certification, whether first or second round, does not prohibit the transfer of land to one's heirs, although there is a requirement that inherited land should be registered and a certificate issued in the name of the new owner. Moreover, it is customary here to divide the family property among all family members, a practice which has been reinforced by the land certification that contains all the names of family members in the certificate, thereby endorsing their inheritance rights. Thus, inheritance rights are one of the most significant causes of the fragmentation of land in the woreda, as well as in South Wollo and many parts of the Amhara region.

The government has invested considerable resources in rural areas like Tehuledere as part of its broad rural development initiative. The development officials in the woreda,[6] who were interviewed for this study, noted that they have not received any climate-change adaptation roadmap to implement, although the development and public support programmes they are carrying out with farmers have several components which are proving useful in enhancing farmers' resilience and climate adaptation efforts. Rural development initiatives in general are meant to address what policy planners believe are the three major problems facing many rural communities, namely endemic food insecurity, high rates of natural resource loss, and lack of or inadequate access to resources. The specific measures implemented on the ground include the distribution of improved seeds, access to chemical fertilizers and agro-chemicals, the construction of irrigation schemes, and the promotion of improved cultivation methods, such as row planting (see Pankhurst and Dom, 2019).

Despite the considerable investment in public programmes, the core problems of poverty, joblessness and food insecurity continue to be high and widespread, and this is for many reasons. To begin with, one of the inherent shortcomings of the implementation of public programs is that it is all too often based on a top-down approach. Public agents tasked with programme implementation are often unprepared to recognize the value of farmers' local knowledge and practical experience, instead being keen to have farmers replace their customary practices with what are considered modern ideas, which the agents themselves have received from higher authorities in their agency. Some of the farmers who were interviewed for this research were unhappy about some elements of the extension programme they were requested to adopt, though they were not, for understandable reasons, willing to talk about it to our research team. Local knowledge and practice are essential for the uptake and sustainability of development as well as climate-change initiatives, which should inform policy planning and implementation. This has not been the case very often in Tehuledere. Moreover, all the development agents interviewed for this study confirmed that, despite the widespread adoption of khat in the woreda, and despite its importance as a source of revenue for the local government, there is no specific extension package for the khat sector, nor are rural agents expected to engage with khat growers. There is a strongly held view among the local authorities that khat consumption is addictive, that it poses a health risk and that it is especially harmful to young people and their future. Another shortcoming that plagues public support programmes is the inadequacy of their implementation. A third factor is that local government is unable to meet the growing demand for support services from both rural and urban communities. Both technical inadequacy and the failure of demand management are the result of a shortage of resources, whether financial, technical or human. Many of the programmes under discussion are funded by the donor community, and there are limitations on accessing funds, as well as accountability on the part of the government on how and for what purpose the funds provided have been used.

Khat cultivation in Tehuledere

The cultivation of khat (*Catha edulis*) is widespread among small farmers in the Tehuledere district; it is found in particular in the semi-dry lowlands and wet mid-altitudes. One particular variety from here, called 'galessa', was at one time considered to be of high quality and was, until recently, in great demand in the khat markets of South and North Wollo and the Afar Region. Khat is a perennial crop and, once established, it can be a source of steady income over many years, with harvests twice or even three times a year. The plant can grow into a tall tree, reaching 25 metres or more; it is not uncommon to see plots in the woreda containing a few khat trees that are 25 years old or older. The valued part of the plant, the leaves, can nowadays command a very high price in the market, especially if they are very fresh and sourced from high-quality 'brands'. The khat plant can tolerate high temperatures and drought. It will go into 'hibernation', as it were, in long dry spells, but will not die, quickly recovering if provided with even a small amount of water, and ready to harvest within 20 days, according to our informants. This makes it a suitable crop in the kind of conditions characterized by rainfall irregularity, as will be described later on in this chapter. All these factors have greatly added to the popularity of the plant, contributing to its rapid spread across the country, as well as in South Wollo as a whole in the last three to four decades. According to informants in the woreda land office, khat now covers 5 per cent of the land in the dry lowlands, 50 per cent in moist 'weina dega' or mid-altitudes, and 20 per cent in the 'dega' or colder and higher altitudes. Khat, according to them, is replacing food and fruit crops at a rapid pace.

Khat is the most expensive crop in the country, and prices for it have continued to rise over several decades, so much so that at present the price of quality khat leaves is much higher than that of coffee, though khat prices are more volatile than for most other crops, changing rapidly depending on the season, quality and even time of day. However, khat is also a highly

Table 6.2: Exports of selected commodities (in US$ millions)

Export items	2018/2019	2019/2020	2020/2021
Coffee	764.1 (28.7%)	855.9 (28.6%)	909.4 (25.1%)
Oilseeds	387.8 (14.5%)	345.0 (11.5%)	335.5 (9.3%)
Flowers	256.6 (9.6%)	422.3 (14.1%)	470.6 (13.0%)
Khat	303.6 (11.4%)	324.4 (10.9%)	402.5 (11.1%)
Pulses	272.3 (10.2%)	234.8 (7.9%)	233.8 (6.5%)

Note: The figures in brackets are the share of each item in total exports for each year.
Source: National Bank of Ethiopia, 2022: Annual Report 2020/2021

perishable product: the 'shelf life' of the picked leaves is about 24 hours or less and delivering them to consumers within this timeframe is one of the most challenging aspects of khat marketing. The khat industry also provides a good deal of employment to young people, especially in rural and semi-rural communities, and has contributed to increased economic activity in small towns in khat-growing areas. Khat is also in great demand in many countries in the Middle East and has consequently become a valuable export commodity for the country, bringing in much-needed hard currency for the government. The data in Table 6.2 show that khat exports made up, on average, more than 11 per cent of the value of all exports annually.

The argument

There is an ongoing debate on khat and its widespread cultivation and consumption in the country (Asnake and Zerihun, 2017).[7] Advocates argue that khat production provides high levels of household income to small farmers, offers considerable rural employment, can serve as an important tool for poverty reduction in the countryside and, as we noted earlier, earns the government valuable hard currency (Ezekiel, 2009; Gessesse, 2013). On the other hand, as its significance has increased in the country, it has aroused concern in many other countries and has become a prohibited substance in Europe and elsewhere (Cochrane, 2016). However, this has not harmed the domestic and export market, as khat continues to be in high demand in the Middle East, which remains a major market for Ethiopia and East African countries that also grow it. In contrast, those opposed to khat argue that the detrimental effect of the cultivation and consumption of the plant outweighs the benefits that are claimed for it. Government officials and development officials in Tehuledere, who were interviewed for this study, stressed their concern that the expansion of khat cultivation is taking place at the expense of food crops, as more and more farmland is converted into khat groves. They note that as the woreda, as well as South Wollo Zone, has frequently been blighted by severe food shortages and famines, the priority should be given to improving food production and nutritional security. Another objection to the khat industry is based on the grounds of personal health and wellbeing, including the harmful impact khat has on young people. Opponents argue that khat has addictive properties and that regular consumption leads to impaired personal health, especially mental health, with harmful consequences on one's work and education, family and social wellbeing (Solomon, 2017; Yeraswork, 2017).

However, in this chapter, it is argued that khat cultivation could just as well be seen within the framework of small farmers' own adaptation strategies to climate change. Our findings in Tehuledere show: (a) that farmers are aware of climate change and that it is affecting their lives; (b) that climate change causes a high degree of uncertainty and insecurity among people because nature and natural processes have now become unpredictable; and (c) that small farmers have devised their own ways of meeting the challenge of climate change by means of what may be described as indigenous or home-grown adaptation strategies. These strategies place khat at the centre of a diverse interplay of farming activities aimed at reducing the risk environment and softening the impact of any localized adversities that may arise. Such strategies were in place well before the government's response programme to climate change was introduced in rural areas, but they have since benefited from them, making them more diverse and more robust, not least because of the provision of improved seed varieties and the construction of some irrigation schemes.

Adaptation strategies of small farmers

Asked to name the variety of adaptive measures in which they had engaged, the farmers in our interview cited the following: planting improved varieties when available (fast-growing and drought-resistant ones are preferred); shifting from cereals to vegetables (chili peppers, onions, potatoes), and from vegetables to fruit trees (mangoes, oranges, avocadoes) depending on market conditions and the availability of improved varieties; using irrigation where available or, if not, pumping water they have collected using generators on their own initiative; intensification of labour; and migration. However, underlying all these efforts and serving as an effective anchor is khat cultivation, which growers said they rely on as a source of income for the family.

The adaptive measures selected by families have implications for household resources and require frequent decisions affecting land use, cropping plans, family labour and household consumption. Managing the demands of adaptation on the one hand and household resources on the other is often a difficult balancing act in which success depends on sound and risk-averse decisions on the one hand, and quick responses on the other.

Climate change

Small farmers in Tehuledere are acutely aware that changes of great consequence are taking place in the natural world whose pervasive impact on their lives and livelihoods they experience on a daily basis. Some of the changes are obvious, while others are less noticeable. Among the first, the most consequential and those our respondents cited most often are changes in rainfall and temperature patterns. The rains, they say, have now become more erratic and unpredictable with regard to their timing, duration and volume, and there is now a longer period of high temperatures and dryness. Farmers are dependent on nature that is predictable, as well as on reliable access to natural resources that are ample enough to enable season-based farming activities to be carried out. Climate change has given rise to conditions that are contrary to these: what we have instead is rainfall variability, rising heat levels and longer dry spells.[8] But farmers also recognize that the natural changes that take place around them have brought about significant consequences that require change and adjustment. These consequences arise from shifts taking place in the seasons (notably the agricultural seasons) and in agro-ecology (ecosystem changes) requiring adaptive changes in farming practices.

A brief note on the agricultural seasons and agro-ecology zones defining livelihoods in Tehuledere and northeast Ethiopia in general is relevant here. There are two main farming seasons during the year in the woreda: the 'belg' and 'meher' seasons. In the belg season, the rains are expected to fall from March to May, and the harvest takes place from June to August, while in the meher season the rains are supposed to extend from June to September, with harvesting from November to January. There are some variations in different localities, but this is broadly relevant for Tehuledere and many areas in South Wollo. The meher season is the most important season for the country's rural economy, for it is in this season that the greater proportions of the country's food and other crops are produced. On the other hand, while its contribution to total agricultural output is comparatively small, the belg season plays an important role, providing food resources during the lean season and pasture for animals, and enabling farmers to be better prepared for the main season. Moreover, there are some farming communities in the dega zones of the country (for more on this, see the discussion later on in the chapter) which depend almost solely on belg season production.

With regard to agro-ecology, the following general zonation scheme is accepted as standard (MoA, 1998):

Bereha (desert): hot, dry lowlands; elevation less than 500 metres above sea level (asl).
Kolla: semi-dry lowlands, 500–1,500 m asl (sorghum, cowpeas).

Weina-degga: 1,500–2,300 m asl (most cereals and legumes).
Dega: 2,300–3,200 m asl (barley, wheat and highland oilseeds).
Wurch: 3,200 m and above (some barley).

The weina-degga, the moist mid-altitude zone of the country, is where most of the rural population lives and most of the country's food crops are produced. According to information from the Amhara Agriculture Bureau, 72 per cent of the farmed land in Tehuledere lies in this agro-ecological category, with kolla and dega zones making up 15 and 13 per cent respectively. Khat has traditionally been cultivated in the weina-dega and upper parts of the kolla, but, as we shall see, this is now changing due to climate change.

Many of the farmers interviewed for this study reported that a gradual shift from a biannual to an annual agricultural season is taking place at present. They all said that the availability of rain in the belg season was no longer dependable and that they had given up belg-season farm work for a while. The following are some examples of the views of respondents on climate change as they are experiencing it:

> At present the rivers have low water levels … Springs and rivers have reduced water levels. … Since 2000 [Ethiopian Calendar (EC); 2007 Gregorian Calendar (GC)] the belg season has been on the verge of disappearing. (A 55-year-old respondent in Hitecha kebele)

> This year, for example, there is no belg because it has not rained since the wet months (the kiremt) … Since 2002 [EC; 2009 GC] due to rainfall variance the belg no longer occurs. As a result, rural people have become fearful … Because it is not raining at the expected times, people are worried. (A 45-year-old respondent in Hitecha kebele)

> Since the coming [to power] of the EPRDF, the climate has become erratic. Our area was weina dega, but now it has become kolla. What was kolla has turned into the Sahara Desert [sic]. What was dega has become weina dega. The rain does not fall at the expected times … there is a high degree of variability. (A respondent, 65–70 years old, in Hitecha kebele)

The vocabulary of climate change among our respondents reflects their lived experience: it is the language of farming, of living and working with nature and depending on it for their livelihoods, as well as for their future. Climate change means things are in flux, there is inconsistency, and hence uncertainty and insecurity. There is a shift in the seasons and agro-ecologies: what was belg is disappearing, and rural dwellers can no longer depend on a second farming season. The distinguishing climatic characteristics of the agro-ecological zones have changed, which means that crops that used to grow in the kolla areas, for example, are now growing in the weina-dega areas. Khat is now being grown at the lower levels of the dega zone, whereas previously it was grown in the weyna-degga and upper levels of qolla zones only. In other words, we are witnessing the upward movement of crops and plants as a consequence of the shifting seasons and agro-ecologies, and farmers are experiencing it in their daily lives. Native seed varieties are losing their value and, according to one respondent, are in the process of disappearing, which may have unforeseen consequences in the future.

Khat plots

The balancing act between adaptation and household resources noted earlier is reflected in the decisions of households to allocate land for khat cultivation. Such is the severe shortage of land that the allocation of land to grow a crop has to be thought through very carefully. According to the March 2021 GCM Household Survey, the mean size of land plots in the woreda is 0.68 hectares.

> In the past we were five in the family; each one subsequently took a 'slice', and land became very scarce. What used to serve five of us before has now been subdivided among us. On the land I received I support six family members. On this land I decided to give up crop rotation and concentrate on one [food] crop. (A 55-year-old farmer in Hitecha kebele)

Nevertheless, in most cases, farmers allot one plot for khat and another plot or plots for food crops, and quite often the khat plot is in the backyard or very close to the homestead. According to the same GCM Household Survey, which was conducted in three selected kebeles, 33.6 per cent of those interviewed said they had planted khat on their backyard plots, 6.6 per cent on their field plots and 7.1 per cent on both, for a total of over 46 per cent (see Table 6.3).

These figures do not provide an accurate picture, as the number of households cultivating khat is much higher. The reason for the low figure may lie in the selection of the kebeles for the survey. Seglen kebele has very little khat cultivation because most of the land here is located in the hot, dry kolla zone, which is not suitable for khat. The GCM Household Survey indeed shows that 92 per cent of respondents there said that they do not cultivate khat. To arrive at a better picture, it may be best to look at the findings for Hitecha kebele, the other kebele in the survey that has a high rate of khat growing. Here, 60 per cent said they have planted khat

Table 6.3: Khat growing in selected kebeles in Tehuledere (percentage of growers)

Plots used	Bededo	Hitecha	Seglen	Total
Backyard plot	33.1	60.3	4.4	32.6
Field plot	5.9	11.8	2.2	6.6
Backyard and field	2.2	17.6	1.5	7.1
No khat plot	58.8	10.3	91.9	53.7

Source: GCM Household Survey 2021

in their backyard, 12 per cent on their field plots and 18 per cent on both, which means that 90 per cent of respondents have khat plots of one sort or another. In Bededo, where 41 per cent of respondents said they have khat plots, 33 per cent said it is located in their backyard. The question now is: given the fact that khat prices are very high and continuously growing, why do most households choose to limit its cultivation to a small plot often in their backyard and why, indeed, does specialization in khat growing or mono-cropping appears to be quite limited?[9]

The farmers interviewed for this study gave a variety of reasons why they have not planted khat on all the land in their possession, and why the common approach is to practise a prudent form of mixed cultivation in which one plot is devoted to khat and other plots to the production of food crops. For many, the concern for food security cannot but be a top priority. Tehuledere and South and North Wollo (and, indeed, the whole of the northeast of the country) have experienced frequent food shortages, including catastrophic famines, over many years, and many of our respondents may have experienced one or two tragedies of this sort in their lifetime. The famines of 1974/1975 and 1984/1985, both of which devastated Wollo, occurred 46 and 36 years ago respectively.

Several reasons were given for why respondents had not used all their land to plant khat: one common reply was it was not advisable to do so, that one should have something to fall back on should your khat be damaged or lost due to disease or accidents of one sort of another, or if there is shortage of food in the market. Others pointed out the high amounts of labour required for khat cultivation, even after the khat plant has been established and is ready to be picked. Khat, farmers noted, requires constant care and attention: the land around the plant will frequently need to be hoed, the plant itself pruned and trimmed, provided with compost and water,[10] and guarded from thieves and other miscreants. Many families do not have sufficient labour power to undertake such tasks on a regular basis and thus are unable to cope. Many observed that khat plots further away from the homestead will not be given equal care and attention, which is why the homestead plot is often the most preferred for growing it. Respondents were not favourably disposed to employing hired labour for khat work, as it will most often prove to be more of a liability than a benefit because they lack the special skill needed for khat management. The other advantages of having khat growing in one's backyard are for ease of care and management on the one hand, and to protect against thieves and pilferers on the other. Khat farms are not infrequently the target of thieves, some of whom are khat addicts from the neighbourhood who steal to satisfy their craving, while others consist of gangs who come from urban areas to steal for monetary gain.[11]

We have discussed how climate change has induced shifts in many aspects of rural livelihoods: shifts in crop choices and farming practices, in the seasons, and agro-ecologies. Another kind of induced shift that is relevant in the present context, though somewhat different in nature, is the shift that has occurred in khat cultivation, this time through a shift from the private sphere to the public, from cultivation for private use to cultivation for commercial gain. Market-driven and widespread khat cultivation is of recent origin in Tehuledere, but this is not to say that the khat plant was unknown in the woreda. Some of the more knowledgeable respondents in our interviews stated that there were a few individuals in their community who had planted small stands of khat during the time of Haile Selassie, those individuals often being religious, while community leaders occasionally held khat-chewing sessions as part of religious practice and social custom. Khat was rarely sold in the market, any exchange occurring privately in the form of gift giving, and the number of chewers was small, often restricted to religious leaders and older and more respected individuals in the community. This all came to an end, we were informed, when khat became a public good, sold in the market for ever-increasing prices, 'exported' to faraway cities and regions, with the number of chewers growing in number and diversity: people of all ages, from rural as well as urban areas, and from many diverse social backgrounds. Many of our respondents place this dramatic shift as beginning in the early years of the 1990s, though some push it back to the late 1980s. According to one of our respondents in Hitecha kebele: 'There were some stands of khat under Haile Selassie, but khat was planted on field plots in great numbers in the woreda from 1983 [EC] onwards [that is, 1990/1991 GC].' He went on to say that as khat cultivation spread, so did khat chewing. This shift from the private to the commercial brought with it greater knowledge of the khat plant and its varieties, as well as greater skill in its management.

It is sometimes argued in the literature that khat's other attraction for small farmers is that it does not require much labour; this was also the view of a few respondents in our interviews. But this is misleading, and for many of our respondents, growing the plant is laborious and time-consuming, as it requires constant care and attention, as already noted earlier. One point to add here is that an important task of keeping the plant in good condition is the application of compost (not chemical fertilizer) on a regular basis, but making compost is also hard and laborious.

Conclusion: Is the khat-based balancing act sustainable?

While the khat-centred adaptation strategy has provided a degree of security to farming households that have adopted it successfully, shielding them from some of the worst effects of climate change, there is cause for concern that this outcome may not last for long. The following are some of the reasons why one cannot but adopt a less optimistic view of its future prospects.

The challenges of climate change, and the consequent insecurities of rural livelihoods, are more likely to be aggravated, not reduced, in the near future (IPCC, 2007, 2022). The IPCC is pessimistic about the effectiveness of local adaptation measures in the face of increasingly severe climate-change impacts and, in particular, extreme weather events (IPCC, 2022). It warns that although 'many early impacts of climate change can be effectively addressed through adaptation, the options for successful adaptation diminish and the associated costs increase with increasing climate change' (IPCC, 2007: 69).

In addition to more severe climate-change impacts, one has to factor in increasing demographic pressure: despite cases of out-migration, rural Tehuledere consists in large measure of an immobile society that is strongly attached to the land; this was confirmed by the findings of the GCM Household Survey noted previously. Indeed, some of our respondents were acutely aware of the shortage of not just farmland but also of 'living space' in their communities, as well as being cognisant that migration, both domestic and international, is becoming less and less viable.

It is imperative that indigenous adaptation is actively complemented by planned adaptation if smallholder agriculture is to be viable in the face of the challenges it will be confronted with in the future. However, there is not much evidence that such integration is actively being sought or promoted at present. The government does not provide any support to the khat sector, nor is there any evidence that it is willing to engage with it, whether at the level of cultivation, commercialization or consumption (Cochrane and Girma, 2017; Selam and Birtukan, 2017). Indeed, many public servants in Tehuledere woreda believe there should be a campaign against khat cultivation.

Notes
[1] This work is based on interviews conducted in April 2022 with several small farmers in Tehuledere, one of GCM's research districts in Ethiopia. I would like to thank Mulugeta Tesfaye Teshome of Wollo University, who undertook the interviews and gathered additional information for me, for the excellent job he has done under the circumstances.

We had planned to run the interviews in three of the kebeles (subdistricts) that had been involved in the GCM Household Survey of 2021, but due to security concerns, we had to abandon our plans for one of them, Seglen kebele.
2   The GCM team made two exploratory visits to Tehuledere to gather information, one in September 2019, the other in March 2020, soon after COVID-19 was reported in the country. On both occasions we talked to officials of the *woreda* administration (in the offices of agriculture, land and environment, labour and social affairs, education, health and the planning commission); we had an extended interview with the Mayor of Haiq town, the district capital. We also talked to a few Haiq residents we met casually, and sampled the coffee served in the 'sidewalk' cafés run by young women in both Haiq and Dessie where we stayed. This section is based on the information we gathered, as well as our observations during these two visits.
3   It should be noted that the data collection took place shortly before the conflict in Tigray.
4   See Asnake and Zerihun (2017) and the papers in that work. See also Gessesse (2013) and Tesfa (2017).
5   See the CSA's population projections report (2013) information for the whole country, as well as for South Wollo.
6   These are *woreda*- and *kebele*-based agricultural experts, natural resource and cropping specialists, extension agents and land-tenure experts; they all work with farmers.
7   For a general discussion of khat in Ethiopia, see the papers in Cochrane (2016); Asnake and Zerihun (2017); Beyene et al (2017); Cochrane and Girma (2017); Girma (2017); Kandari et al (2014); Tesfa (2017); Yeraswork (2017); and Petros (2020).
8   These climate change observations are consistent with the GIS findings provided in Chapter 3, which found that maximum temperatures have been increasing annually since 1983, while the rainfall regime has shown a marked degree of variability in the same period.
9   Limiting khat growing to small plots is also practised in other parts of the country; see Gessesse (2013) and Beyene et al (2017). In the former case, khat growers expanded their food crop plots.
10 Many respondents stated that irrigation water is not best for khat; rainwater or hand watering is said to be better. Also, there is preference for compost rather than chemical fertilizers.
11 One strong reason for backyard plots (not mentioned by respondents) is that ownership of such plots are unlikely to be contested and therefore are more secure than field plots.

References

Bassett, T. and Crummey, D. (eds) (2003) *African Savannas: Global Narratives and Local Knowledge of Environmental Change*, Oxford: James Currey.

Beyene Wondafrash Ademe et al (2017) 'Khat production and consumption; its implications on land area used for production and crop variety production among rural households of Ethiopia', *Journal of Food Security*, 5(4): 148–154.

Cochrane, L. (2016) 'Legal harvest and illegal trade: trends, challenges and options in khat production in Ethiopia', *International Journal of Drug Policy*, 30: 27–34.

Cochrane, L. and Negash, G. (2017) 'Developing Policy in contested space: khat in Ethiopia', in K. Asnake and M. Zerihun (eds) *Multiple Faces of Khat*, Addis Ababa: Forum for Social Studies, pp 143–162.

Cochrane, L. and Singh, R. (2017) *Climate Services for Resilience: the Changing Roles of NGOs in Ethiopia*, BRACED.

CSA (Central Statistical Agency) (2013) *Population Projections of Ethiopia 2007–2037*, Addis Ababa: CSA.

CSA (2021) *Report on Area and Production of Major Crops (Private Peasant Holdings, Meher Season)*, Addis Ababa: CSA.

Dessalegn, R. (2003) 'Littering the Landscape. Environmental Policy in Northeast Ethiopia', in T. Bassett and D. Crummey (eds) *African Savannas: Global Narratives and Local Knowledge of Environmental Change*, Oxford: James Currey, pp 205–224.

Dessalegn, R. (2009) *The Peasant and the State. Studies in Agrarian Change in Ethiopia 1950s – 2000s*, Addis Ababa: Addis Ababa University Press.

Dessalegn, R. (2013) 'Food Security and Safety Nets: Assessment and Challenges'. In Dessalegn, R., Pankhurst, A. and van Uffelen J.G. (eds), *Food Security, Safety Nets and Social Protection in Ethiopia*, Addis Ababa: Forum for Social Studies. pp 113–146.

EAS (Ethiopian Academy of Sciences) (2015) *Ethiopian Panel on Climate Change. First Assessment Report*, vols I–VII, Addis Ababa: Ethiopian Academy of Sciences.

FDRE (Federal Democratic Republic of Ethiopia) (2010a) *Nationally Appropriate Mitigation Action (NAMA- Ethiopia)*, Addis Ababa: Environmental Protection Authority.

FDRE (2010b) *Growth and Transformation Plan I 2011–2015*, Addis Ababa: Ministry of Finance and Economic Development.

FDRE (2011a) *Ethiopia's Climate-Resilient Green Economy* Green Economy Strategy, November, Addis Ababa.

FDRE (2011b) *Ethiopia's Vision for a Climate Resilient Green Economy*, Addis Ababa.

FDRE (2011c) *Ethiopia's Programme of Adaptation to Climate Change (EPACC)*, Environmental Protection Authority, Addis Ababa.

FDRE (2012) *Chat Excise Tax Proclamation. Proclamation 767/2012*, Federal Negarity Gazeta, August, Addis Ababa.

FDRE (2020) *Ten Years Development Plan: A Pathway to Prosperity, 2021–2030*, Planning and Development Commission, Addis Ababa.

Gebissa, E. (2009) 'Scourge of life or an economic lifeline? Public discourses on *khat* (*Catha edulis*) in Ethiopia', *Substance Use & Misuse*, 43(6): 784–802.

Gessesse, D. (2013) *Favouring a Demonised Plant: Khat and Ethiopian Smallholder Enterprise*. Uppsala: Nordiska Afrikainstitutet.

IPCC (Intergovernmental Panel on Climate Change) (2007) *Climate Change 2007: Impacts, Adaptation and Vulnerability*, Cambridge: Cambridge University Press.

IPCC (2022) *Climate Change 2022: Impacts, Adaptation, and Vulnerability. Summary for Policy Makers*. Contribution of Working Group II to the Sixth Assessment Report of the Intergovernmental Panel on Climate Change. Cambridge: Cambridge University Press.

Kefale, A. and Mohammed, Z. (eds) (2017) *Multiple Faces of Khat*, Addis Ababa: Forum for Social Studies.

MoA (Ministry of Agriculture) (1998) *Agro-ecological Zones of Ethiopia*, Natural Resource Management and Regulatory Department, Addis Ababa.

National Bank of Ethiopia (2022) *Annual Report 2020/21*, Addis Ababa.

Negash, G. (2017) *The Education of Children Entangled in Khat Trade in Ethiopia. The Case of Two Khat Market Centers*, Addis Ababa: Forum for Social Studies.

Pankhurst, A. and Dom, C. (2019) *Rural Ethiopia in Transition: Selected Discussion Briefs, 2018*, Addis Ababa: WIDE.

Petros Terefe Tolcha (2020) 'Khat marketing and its export performance in the Ethiopian economy', *Science Research*, 8(4): 90–97.

Selam Gebrehiwot. and Birtukan Haile (2017) 'Khat and the need for legislative reform in Ethiopia', in K. Asnake and M. Zerihun (eds) *Multiple Faces of Khat*, Addis Ababa: Forum for Social Studies, pp 125–141.

Teferra, S. (2017) 'Impact of khat on health with emphasis on mental health', in K. Asnake and M. Zerihun (eds) *Multiple Faces of Khat*, Addis Ababa: Forum for Social Studies, pp 109–123.

Tesfa Binalfew (2017) 'The expansion of production, marketing and consumption of chat in Ethiopia', *International Journal of Research in Agriculture and Forestry*, 4(3): 16–26.

World Bank (2010a) *Economics of Adaptation to Climate Change. Ethiopia*, Washington DC: World Bank.

World Bank (2010b) *Economics of Adaptation to Climate Change: Synthesis Report*, Washington, DC: World Bank.

Yeraswork, A. (2017) *The Khat Conundrum in Ethiopia: Socioeconomic Impact and Policy Directions*, Addis Ababa: Forum for Social Studies.

7

'They Are Supposed to Stay Home': Examining the Politics of Adaptation and Climate (Im)mobility in Northern Ghana

Lily Salloum Lindegaard, Francis Xavier Jarawura and Nauja Kleist

Introduction

Policy approaches to climate-related mobilities are emphasizing in situ adaptation – efforts to adapt in place – to limit migration as a response to climate-related stress. Such approaches reflect portrayals of climate-related mobility as negative responses to climate change, as for instance in narratives of climate refugees, migration as a failure to adapt, or migration as maladaptation. Yet an alternative body of research on migration questions whether this is a relevant or desirable approach and for whom. This literature emphasizes migration's central role in existing livelihood and development strategies more generally (Abdul-Korah, 2011; Teye et al, 2017; Setrana and Kleist, 2022), for example, through remittances, as well as migration as a potential adaptive and coping strategy for the migrant themselves and often for their households and social networks in their places of origin as well (see, for example, van der Geest, 2011). Discussions continue of the 'positive' and 'negative' framings of climate-related migration. This chapter contributes to these discussions by considering how such framings are mobilized in policy and practice and with what implications for whom.

In line with this, the chapter provides a critical analysis of a policy initiative to limit migration in a climate-affected in northern Ghana through in situ irrigation as adaptation. It examines the politics and outcomes of this initiative, with emphasis on exploring the differentiated perspectives among actors within and outside the affected community, and how these are mobilized and contested. Specifically, it considers the case of irrigation to provide dry-season livelihoods in two small villages in the Upper West region of Ghana where the majority of the population is dependent on rain-fed agriculture as their main livelihood. Poverty in the region is among the highest in Ghana (GSS, 2018), and the region lacks economic and infrastructural development that could support alternative livelihood opportunities. This is linked to historical development inequalities between the north and south of Ghana, which now influence climate change impacts and mobility responses, specifically current north-south migration, as will be discussed in more detail later on (Gravesen et al, 2020). Current irrigation interventions attempt to counter these trends by seeking to bolster rural livelihoods, reduce poverty and provide adaptation to increasing climate variability linked to climate change.

To understand the politics and outcomes of in situ adaptation as a tool to manage mobility, the chapter puts forward a novel analytical approach bridging two literatures: climate mobilities and politics of adaptation literatures. Together, these provide insight into how agendas and perspectives of different actors shape climate interventions as well as migration practices and perceptions, including those of the affected communities. This supports analysis of the political and social rationales informing adaptation projects' inception, implementation and effects, going beyond the narrow confines of adaptation to link to wider political aims and processes. Through our analysis, we find that this joint approach embeds adaptation and climate mobility in a broader context of power and politics, where differentiated interests, influence and impacts for different actors and groups emerge.

The following section briefly explores literatures on the politics of adaptation and climate mobilities, and how these can be linked in a joint analytical approach. The cases of irrigation are then presented, including the historical development and current socioecological context of the case areas; the political processes through which the irrigation initiatives were framed and implemented; local contestations around migration, particularly in relation to generational differences; and the mixed outcomes of the irrigation schemes, including for climate-related mobility. We then offer reflections on the utility of the joint climate mobilities/politics of adaptation approach. Finally, we present key findings and implications for research and policy going forward, particularly to what extent and for whom in situ adaptations can address mobility aspirations and adaptation needs.

Overall, we find that policy efforts to limit climate-related mobility through in situ adaptation programmes will likely be challenging and have mixed results, including some detrimental outcomes. While they may provide alternative livelihoods for some, they are unable to address the many and diverse risks posed by climate change. In addition, they enter into a setting of contested mobility practices and will likely struggle to address competing aspirations and needs across community groups.

The politics of adaptation and climate mobility

Adaptation literature has increasingly highlighted the highly political nature of adaptation. Critical scholars especially have reframed adaptation from a fairly technical undertaking focused on addressing external climate change impacts (Taylor, 2015; Mikulewicz, 2018, 2020) to 'a socio-political process that mediates how individuals and collectives deal with multiple and concurrent environmental and social changes' (Eriksen et al, 2015: 523; Ojha et al, 2016). This reframing has required a new set of analytical approaches and concepts to study the politics and power that researchers argue shape the need to adapt to climate change – that is, differentiated vulnerability and exposure to climate variability and shock, the focus and nature of adaptation interventions themselves (Eriksen et al, 2015), and differentiated responses to and outcomes of these interventions (Eriksen et al, 2021). Various researchers have since drawn on diverse disciplines, literatures and concepts to explore this new framing of adaptation. This includes, among others, subjectivity, knowledges and authority (Eriksen et al, 2015), political representation (Nagoda and Nightingale, 2017; Nightingale, 2017), problematics of government and political rationalities (Lindegaard, 2018), and biopolitics (Bettini, 2017a).

A similar political framing is gaining traction in the study of migration and mobility in response to climate variability and change. Debates on climate-related mobility have long been entangled with political questions and framings, for instance, in early debates on 'climate refugees' imbued with questions of justice and responsibility (Gemenne, 2015; Bettini et al, 2017). In recent years, these debates have been particularly evident in 'climate mobilities' literature, with nuanced analysis of the complex movements of people in settings affected by climate change, including attention to underlying socioecological and power relations (Boas et al, 2018, Weigel et al, 2019). This body of literature is increasingly engaging with the governance literature that has shaped critical adaptation scholarship. This is evident in the presentation of new research agendas on the situated politics of climate mobilities (Boas et al, 2022) and the governance of climate mobility (Lindegaard et al, 2024). Overall, these contributions draw attention to the relationality and differentiality of (im)mobilities (Weigel et al, 2019), and the dynamics of access and representation that shape mobilities and how these are governed (Lindegaard et al, 2024).

In this chapter, we both draw on and link these two literatures to study the intertwined politics of adaptation and of climate mobilities. Specifically, we draw on conceptualizations of both adaptation and climate mobilities as sociopolitical processes, entangled in specific socioecological contexts and histories (Eriksen et al, 2015; Lindegaard et al, 2018; Weigel et al, 2019; Jarawura et al, 2024). In the first instance, this draws attention to how differentiated vulnerabilities and mobilities in the case areas have been produced, through particular political, social, economic and environmental processes over time, and how various groups and individuals are affected and engage with these (Eriksen et al, 2015; Ribot, 2017; Weigel et al, 2019). In the second, this approach points to the political and power dynamics through which adaptation interventions seeking to shape climate mobility – in this case irrigation – are justified and formulated (Bulkeley and Stripple, 2013; Lindegaard et al, 2024). To study this, we specifically draw on adaptation literature calling attention to the import of power dynamics, animated through access to and representation in decision-making processes (Ojha et al, 2016; Nagoda and Nightingale, 2017). Finally, we draw on literature on implementation of adaptation interventions and differentiated responses and outcomes in practice. With this, we seek to capture potentially negative impacts of the irrigation initiatives, prompted by a growing body of critical literature describing the dissonance between adaptation aims and outcomes and the potential for adaptation efforts to exacerbate or redistribute vulnerability across and within communities (Eriksen et al, 2021).

In line with this analytical approach, the following section will explore the production of precarity and differentiated vulnerability in the Northern Savannah zone of Ghana, where the two case villages are located. It also describes the socioecological context, focusing on the interrelations between agriculture, out-migration and climate change. The chapter then examines the politics of irrigation as in situ adaptation, highlighting the political processes behind particular irrigation schemes in the case study villages, outlining the role of differentiated interests, power and political representation in producing these interventions, and framing them as an alternative to out-migration in a climate-affected area. We then zoom in on intra-community dynamics around these irrigation projects, including practices, perspectives and tensions around migration. Finally, we look at the outcomes of interventions, considering these against the political rationales behind them.

Climate change and migration in Northern Ghana

The Upper West region, where the two case villages are located, is a semi-arid zone of Ghana. It is often studied in relation to climate change in Ghana, as environmental conditions are already challenging and socioeconomic development is low, suggesting greater vulnerability to climate change (see Chapter 5 on climate change and adaptive capacity). Despite these challenges, rain-fed agriculture is the main economic activity in this region, and commonly cultivated crops include groundnuts, maize, yam, Bambara beans, rice and sorghum. However, the agricultural economy has not been enough to lift residents beyond the poverty line, and the region has persistently had the highest reported poverty incidence in Ghana – most recently at 70.9 per cent, in contrast with 2.5 per cent in Greater Accra in the south (GSS, 2018).

These forms of precarity are also evident in the two case villages in the Upper West region, which for reasons of anonymity we refer to here as 'Village 1' and 'Village 2'. Both are typical agrarian communities where people largely depend on rain-fed farming. Village 1 is located in the Jirapa district. Residents mainly engage in agricultural livelihoods. Most cultivation takes place in the single rainfall season, and few households can engage in agriculture during the dry season. Village 2 in the Wa West district has similar characteristics to Village 1.

The persistent poverty and increasing migration from the region are closely linked to historical development policies of the colonial and post-independence periods. The colonial government deliberately limited socioeconomic development in northern Ghana to ensure continuous labour flow from the north to the south, to support revenue-generating mines and export crop plantations. Agricultural (including irrigation), health and educational developments were neglected, with investments instead being channelled into the south (Songsore and Denkabe, 1995, Jarawura et al, 2024). While some efforts were made in the later colonial period, these were limited and often marred by failures in implementation (Gravesen et al, 2020). Current forms of precarity such as poverty and climate risk are thus in many ways informed by historical developments.

Migration has been a key response to environmental stress and shocks over recent decades, as 'a way of dealing with structural environmental scarcity and non-farm opportunities' (van der Geest, 2011: 90). It constitutes an important livelihood in both villages. There are little to no opportunities for agricultural income generation during the long dry season (October–April), and young people in particular look elsewhere in search of jobs in this period. The southern part of Ghana has been the main destination as it possesses greater opportunities for both farm and nonfarm jobs. First, it has a double maxima rainfall regime, which allows for crop cultivation throughout the year, including at plantations. This provides good conditions for seasonal migration for farm work. There have also been permanent migrations out of the villages in search of better ecological conditions. Second, the concentration of extractive industries and the relatively higher levels of urbanization in the south provide a range of nonfarm jobs. The mines in particular have attracted many northern migrants for about a century.[1] They range from official large-scale to smaller and often illegal surface mines, known as 'galamsey'. Youth migration to urban centres has also intensified since the 1980s, where women and girls started joining their male counterparts in looking for menial labour (cf. Mariama and Ardayfio-Schandorf, 2008; Abdul-Korah, 2011).

Though migration is an established livelihood practice in the area, environmental change, livelihood opportunities and mobility are experienced and viewed differently by various actors in Ghana and beyond. Ghanaian government authorities, civil society organizations (CSOs), and also some people in the two villages perceive migration as a potential and sometimes serious problem that needs to be addressed. Youth migration to urban centres is often framed as particularly problematic. There is thus some resonance with policy discourses of local adaptation as a 'solution' to climate change as well as to migration. We will explore this in more detail in relation to irrigation in the next section.

Irrigation schemes: 'centre stage of agricultural development'

Current irrigation infrastructures are framed in terms of present adaptation needs, but are also informed by the case area's socioecological history. We explore this history by examining how, over time, particular forms of knowledge and authority have shaped the area's physical environment and lived realities, and vice versa.

Overall, irrigation development has long been considered an important intervention in supporting crop production, particularly for building climate resilience among farmers in northern Ghana. Since the 1960s, many discourses and political agendas have advocated for irrigation schemes and dams to enable dry-season cultivation (Yaro, 2004), with some major dams and medium irrigation schemes being constructed, such as the Tono and Buntanga irrigation schemes in the Northern region and the Upper East region respectively. The push for irrigation in northern Ghana gained momentum after devastating droughts of the 1980s that, in combination with political crises, caused widespread food insecurity. Irrigation projects were further emphasized after the realization in the 1990s that the World Bank and the International Monetary Fund's (IMF) Economic Recovery Programme (ERP) did little to move the people of northern Ghana out of poverty (Yaro, 2004). The significant gains in macro-economic indicators, diversification into nontraditional exports and improvements in productive inputs were far less beneficial to the north than the south. Among several reasons for this was a focus on export crops that could not be grown in northern Ghana and the failure to adequately consider the unique climate situation of the region (Boafo-Arthur, 1999). Given this situation, several studies and development agencies recommended irrigation development as a key strategy to deal with the single rainfall regime and high rainfall variability.

State policy emphasis on irrigation was further accentuated from the late 2000s when policy makers became increasingly concerned about the threats and impacts of climate change in the area. This is reflected in the priorities of the Savanna Accelerated Development Authority (SADA),[2] a government institution established in 2010 to reduce poverty in the Savannah agro-ecological zone and bridge the development gap between it and the south of the country. In the 2016 SADA Development Masterplan, inadequate access to water is described as dictating 'present economic and social conditions' (SADA, 2016: 6). As a consequence, irrigation and other 'climate-smart practices' are 'at the centre stage of agricultural development in the NSEZ [Northern Savannah Ecological Zone]' and are expected to 'strongly contribute to climate-resilient agriculture for years to come' (SADA, 2016: 6).

Irrigation has thus been expected to reduce vulnerability to erratic and changing climate conditions by helping farmers cultivate crops in the dry season and build resilience by providing an extra source of income (Antwi-Agyei et al, 2014; Fagariba et al, 2018). Irrigation has also been expected to reduce high out-migration. While north-south labour migration was encouraged during the reign of colonialism, government discourse and policies over recent decades have sought to reduce or stem these migratory movements because of the perceived negative consequences in both the north and south. This is explicated in the SADA masterplan that describes the prospects of irrigation: 'The potential impact is in the creation of employment both off-season and dry season, cultural diversity, reduced rural-urban migration, [and] reduced flooding' (SADA, 2016: 52).

Irrigation constitutes a panacea in this document, in other words, an all-encompassing solution expected to create jobs, support culture, reduce poverty and ultimately reduce migration. This emphasis on irrigation coalesced in the so-called 'One Village-One Dam' (1V1D) policy agenda, which was initiated in 2018 as part of a Canadian government-sponsored 'Ghana Agricultural Development Programme' (GADP). As indicated in the 'One Village-One Dam' name, the government indicated it would establish one dam – whether a larger dam, an earth dam or a dug-out – in every village in the northern part of the country to reduce poverty by providing water for both dry-season farming and other livelihood purposes, such as cattle rearing. Like in the SADA masterplan, such dams are explicitly envisioned to bring food and jobs to local areas and reduce north-south migration:

> Another complementary intervention in the agricultural sector is the 'One Village-One Dam' initiative. This initiative is being introduced especially in the Upper West, Upper East and Northern Regions, to ensure availability of water for all-year farming. This is expected to

improve food security and curtail migration from the north to the south in search for jobs during the off-farm season. (MoFA, 2018: 60)

As indicated previously, a variety of different governance actors are involved in the push for irrigation as a panacea solution to climate vulnerability, economic development and livelihoods, food security and out-mobility. This includes the technical authority SADA/NDA, the national government and in this case the Canadian donor in cooperation with the Ghanaian state. Through irrigation, these actors influence management of territories and people and resources within them (Vandergeest and Peluso, 1995), a dynamic emerging clearly in a range of other adaptation efforts, as well as in relation to climate mobility (Funder et al, 2018; Lindegaard, 2018). This political agenda of reducing climate mobility in Ghana is linked to key assumptions that are clear in the SADA and Ministry of Food and Agriculture (MoFA) documents referred to earlier: first, that of problematic mobility stemmed by local irrigation, which is clear in the documents' portrayal of migration as a deviation from the norm or ideal; and, second, that the mechanism through which mobility arises is a simple causal relation, where inadequate dry-season food and livelihoods act as a push and addressing this single mechanism will significantly reduce migration. These portrayals of mobility, the manner by which it arises and the effect irrigation will have on the same are contested in much of the scholarly literature on climate-related mobility, which instead describes it a typical social phenomenon and highlights the complexity of factors that shape it (Boas et al, 2018; Weigel et al, 2019). However, these portrayals constitute particular forms of knowledge promoted by powerful institutions that are able to shape policy and programming for populations on the ground. Under the 1V1D initiative, a range of medium- and small-sized dams have been constructed, but the initiative is far from being fully implemented.

Rather than being a case of externally driven, locally contested interventions (see Lindegaard and Sen, 2022), the 1V1D projects are often seen as highly desirable by diverse subnational actors, though they are also part of complex tensions and contestations. Regions and communities compete over receiving irrigation schemes, lobbying to attract projects for irrigation or simple dams. This was, for instance, the case in Village 1. During a focus group discussion, community elders recounted how they lobbied for the dam in the 1980s to help reduce the incidence of both food insecurity and out-migration. They reached out to their District Assemblyman, who put them in contact with other influential elites. Together, they made a plan which led them to the Regional Minister representing the military government and presented a petition for the dam. Their petition fell in line with the military government's concern about the negative effects of out-migration from the north and was therefore forwarded to the national level for further consideration. A few years later, engineers came to assess the region's irrigation capacity, and the village was found to have a good natural potential for collecting rainwater. This history illustrates how a series of intersecting political agendas by a range of authorities across scales – an Assemblyman, a Regional Minister, national-level authorities, and village elders – catalysed the construction of this particular dam. This exemplifies adaptation as a multiscalar sociopolitical process (Eriksen et al, 2015: 526), rather than as a technomanagerial process, as it is often portrayed. The preceding account also documents how particular political agendas and the knowledge and authority that animate them have come to physically shape the built environment in this area.

Migration contestations: 'bigger dreams' versus 'food for the household'

Once built, the dam and irrigation scheme entered a setting of contestation about migration in the villages themselves. Though migration is an established livelihood practice in the area, environmental change, livelihood opportunities and mobility are experienced and viewed differently both within and across community groups in the two villages. Most pronounced is the divide between older community members and youth. While young men and women continue migrating to the south of the country, older men and women find the opportunity for minimal cultivation in the dry season enables them to stay in the area:

> Many young people need real money and not food so they migrate to the south where there is illegal gold mining and work in the second rain season. Over there cash is paid for their labour and some young men too work in the city as masons, carpenters and casual labourers … We, the older people, are usually OK with very little money so the irrigation farming helps us to stay as we don't lack to eat. (Mixed-gender focus group discussion (FGD), Village 1)

As expressed in this statement, there are pronounced differences in generational perspectives on the effects of the irrigation scheme and on migration more generally. These emerge largely from the acceptance or refusal to seek alternatives to agricultural livelihoods and subsistence farming. The older generations seem content with the gains from irrigation, while the youth envision migration as a pathway to pursue their hopes and aspirations (see, for example, Kleist and Thorsen, 2017). According to older interviewees and FGD participants, youth migration includes the urge to experience life beyond the village as a sign of maturity and an opportunity to freely explore aspects of modern life, such as income autonomy and less social control, also in relation to intimate relationships. For the older generation, such practices are generally seen as a potential social and moral problem. Another issue, according to an elderly man, is that youth migration reduces household labour that could otherwise be used for farming with no or little cost to the heads of household. Hence, youth migration is not only perceived as morally questionable by some, in terms of change of lifestyle and mentality that affects family, but also affects household livelihoods and economies.

Positive aspects of migration were also acknowledged. 'When our people migrate, they are able to help their households better than when they are here because the bring food and money', explains an old woman. This emphasis on remittances is well captured in the academic literature, where female migrants are described as being more devoted remitters – especially to other women – than men (see, for example, Abdul-Korah, 2011; Teye et al, 2017; Setrana and Kleist, 2022). Yet, young men migrate to take care of their families as well, as exemplified by a teenager who had migrated to do 'galamsey' when he was 16 years old to earn money to support his younger siblings' schooling. He was now back and wanted to stay, but highlighted the lack of opportunities for income in the village.

There are thus tensions among villagers regarding the implications and desirability of migration, especially youth migration. From the perspective of some elders, youth are seen as being overly materialistic or ambitious regarding their futures. A focus group participant described young people as having 'bigger dreams. They want to be big people in the future and to have nice electronics, motorcycles and big certificates and jobs'. However, turning such 'bigger dreams' into reality is difficult in the villages and requires temporary or long-term migration to places with higher incomes.

While the youth 'dream bigger dreams', according to the elders, older migrants explain their migration as reflecting a need to secure food and maintain social order. Focus group participants described that migration for older generations is usually not inspired by aspirations for a different future, but driven largely by the need for food and tempered by social relations at home: 'They migrate to find food for the household', one person explained, continuing that 'they are supposed to stay home and watch over their families and take important decisions in the village' (mixed-gender FGD, Village 1). The older generation is expected to maintain a presence at home and attend to family and clan business. However, their migration can be tolerated if it arises out of dire need:

> Those days when there was hunger we knew and understood that poor older men and women would have to migrate but today how can we understand when you can find vegetables to eat from the work in the irrigated farms ... although it is somehow acceptable for the old ones to migrate to rural areas to farm but not to the city to enjoy, we still

prefer them to stay here with us ... But of course, we know that only few people can buy a fence or sleep in the open at night to keep animals away, so that's the situation we face. (Mixed-gender FGD, Village 1)

Focus group interviews agreed with this assertion that rural-rural migration is considered acceptable for older generations; they emphasized that the breakdown in traditional norms and collectiveness is largely a result of the effects of rural-urban migration, particularly the long stays away from home associated with acquisition of habits that are seen as inconsistent with local norms (Lindegaard and Jarawura, 2024).

There is thus a further distinction between rural-urban migration and rural-rural migration where the latter is perceived as more acceptable for the elder generation. For older people, articulated aspirations are linked to the household rather than the individual, with the expectation that migration will come to an end once the endeavour of providing for the household has succeeded. Social order and social reproduction can then be restored with the older people in their proper place. While migration can be accepted, it seems, if it is short term and explicitly linked with collective purposes, the village is the locus for respectable life in such a perspective. This leads some community members, especially the older generation, to call for measures to curb outward youth migration because of its perceived negative implications for household labour, social habits and expectations picked up in the migration destinations, whether cities, mines or elsewhere.

The dam and irrigation interventions thus enter into complex local politics and social relations, where it is clear that mobility remains a contested activity, closely linked to shifting social norms and social reproduction. This involves forms of authority and social institutions and their (re)production and contestation. Just as the irrigation project becomes part of the changing ecological context to be navigated, so too does it become part of this changing (and interrelated) sociopolitical context that individuals and groups navigate and respond to in a variety of ways. Far from the irrigation scheme being an independent adaptation project, it is thus part of adaptation as a 'contested social-political process that mediates how individuals and collectives deal with multiple types of simultaneously occurring environmental and social changes' (Eriksen et al, 2015: 524).

Irrigation outcomes: 'the story is not over just like that'

Even as community members navigate the new socioecological context around the dam and irrigation schemes, outcomes have been mixed and, in some instances, outright negative. The construction of the dam in Village 1 was initiated in the early 1990s, but after constructing the main reservoir and the main canals, the project was halted and never completed. More recently, one of the main canals broke down, and the intervention now provides only limited irrigation, at roughly 50 per cent of its capacity. Because of the incomplete infrastructure and subsequent partial breakdown, farmers who try to grow vegetables and other crops for the dry season have to draw the water physically from the main canals or utilize water pumps run by diesel to water their crops. This has proven to be very laborious and expensive, and reduces profit levels. Many farmers argue that the farms at the dam site are only good for bolstering vegetable consumption in the dry season when the rains are absent. The farms are small and the soils are poor. Furthermore, the farms offer no opportunities for paid labour for potential migrants during the dry season as families employ their own members. Consequently, the youth have lost hope in the irrigation scheme as it has failed to provide the expected improvement in income. A young married man expressed this as follows:

> When they came to upgrade the dam about five years ago, me and my friends were very happy, and we thought that would end our troubles we face working too hard in the south and gaining little. The dangers there too are a lot from snake bites to exploitation. But the government and his NGOs [nongovernmental organizations] only put up a fence and left us like that to manually fetch the water. The soils are also dead so when you grow crops the harvest is poor. This explains why in spite of the existing dam many young people like me still migrate seasonally to the south. (Young male respondent)

This statement succinctly captures the failure of the dam to bolster jobs and incomes through dry-season irrigation. Not only does the poor construction and uncompleted nature of the water system hinder production, but the degraded soils also stand in the way. The incomplete nature is blamed on 'the government and his NGOs', indicating potentially unclear governance structures and responsibility of the dam construction.

Another unanticipated outcome of the dam in Village 1 is isolation of the village during the rainy season, usually from the months of August to October. Due to its incomplete banks, it overflows after heavy rainfall and blocks the main route into and out of the community. Thus, farmers who harvest crops during these months from rain-fed farming usually face difficulties in transporting produce to the market, as narrated by a 47-year-old return migrant:

> I stopped going southwards to look for work when they constructed the dam ... But the story is not just over like that, it is mixed ... I was nearly carried away by the water from the dam when I wanted to cross to Wa market last year. That day I was carrying tomatoes to

sell in the market but unfortunately, I fell and all the produce were carried away by the water, I was able to retrieve only my container. This has been a serious challenge we the residents here face ... and a bigger problem to our women who depend much on trading. (A 47-year-old return migrant)

As the opportunity to go to the city of Wa through the only other route, via a neighbouring village, is no longer traversable by motor vehicle due to a collapsed bridge, many young people have begun to migrate earlier than usual in recent years to escape being trapped in the village for long. The adjustment of migration patterns to avoid involuntary immobility was described as common among those who already engage in seasonal migration as a livelihood strategy, though some noted it also includes young men travelling recreationally 'to have a rest after the farming season'. In both cases, the harvest of farm products is left to those remaining, mostly younger children and middle-aged to older residents.

As in Village 1, the water system of the dam in Village 2 has no major infrastructure. A simple notch allows the water to gently flow down the slope in a gully. From there, farmers draw water manually with containers or by using small-sized pumping machines to irrigate their crops. There was a general consensus in interviews and FGDs that the dam has had little effect on reducing out-migration from the village, though interviewees pointed to differentiated impacts across households. Households able to fence a plot and afford the cost of pumping water from the dam have benefited most and have seen a reduction in out-migration. Conversely, those with less capabilities and poor connections, for example, to NGOs and politicians to gain support, have seen a continuation or rise in out-migration, we were explained. Hence, while access to farmlands in the irrigable area is not a challenge, the ability to invest in fencing and water pumping systems, given the uncompleted nature of the scheme, are constraining enough to prevent some households from benefiting from the dam. Another challenge for dry season farming using water from the dam is strange diseases affecting vegetables. A farmer reported: 'Almost everybody has access to a piece of cultivable land for dry season faming. The main problem is that most of crops die off after some time, especially tomatoes. This has made many people feel lazy to cultivate in the dry season' (40-year-old man, Village 2). Such problems were confirmed by an extension officer in the area, who explained that while the dam provided access to water for some, it was not able to reduce other climate-related risks, for instance, from disease or sudden, heavy rains that flooded farming plots and drowned plants.

The challenges presented by the incomplete and seemingly abandoned irrigation project have contributed to limit livelihood outcomes for households. It reflects now well-established literatures on differentiated adaptation outcomes and maladaptation (Schipper, 2020; Eriksen et al, 2021), as well as increasing attention on how interventions can increase or redistribute vulnerability or increase climate-related risk (Eriksen et al, 2021, IPCC, 2022), especially for marginal populations. In this case, the intervention has resulted in unintended, sometimes perverse outcomes. Out-migration from the area has increased, in some instances as a direct result of the dam. This is not only due to the dam's failure to address the multiple climate-related risk factors facing dry-season agriculture; it is also due to the failed construction and maintenance of the dam and lack of agricultural extension support for dry-season farming, thereby creating new risks for residents.

Reflections: the politics of adaptation and climate mobility

In the previous section, we presented our analysis using a joint climate mobility and politics of adaptation approach. Overall, we find the combination to be analytically productive, providing insight into the interactions between the politics of adaptation and mobility, efforts to shape climate-related mobilities, and mobility aspirations and outcomes. Here, we reflect on what this joint approach has offered our analysis of irrigation and migration in northern Ghana, link to existing literature, and highlight three specific contributions that will be important for research to provide as climate-related mobility is increasingly the focus of policy and intervention.

First, the joint approach enables insight into social and political rationales through which various actors engage with climate mobility as a focus for policy and intervention. Mobilities literature has done much to illuminate the complexities of climate-related mobilities, essential knowledge for successful interventions able to support adaptive forms of mobility. This includes studies on the complex drivers of migration under adverse circumstances (Black et al, 2013; van Hear et al, 2017), differentiated hopes in and aspirations for migration (Carling, 2002; Kleist and Thorsen, 2017; de Haas, 2021), and the importance of social networks and migration infrastructures (Lin et al, 2017; Kleist and Bjarnesen, 2019). The important next step for research is not only to identify these rationales and hold them up against the latest research findings, supporting critical reflection, but also to understand the mechanisms through which these are mobilized and come to drive or affect specific policies and interventions. In our analysis, we illustrate this by tracing the cross-scalar processes driving particular irrigation interventions and the formal and informal political processes through which this occurs. This includes the differential power of various actors, which is embodied in disparate access to and representation within political processes, which ultimately have direct implications for their livelihoods, water access and mobility options through the dam and irrigation schemes. This suggests the relevance of climate mobility research with explicit attention to political mechanisms of authority, access and representation.

Second, the joint approach provides insight into how adaptation interventions interact with existing mobility patterns and the contested and changing socioecological contexts of which they are part. In our analysis, contestations around climate mobility among actors and across scales emerged very clearly. Such contestations over the modes and implications of migration are not unique to northern Ghana or West Africa (Pickbourn, 2011; Teye et al, 2017; Setrana and Kleist, 2022). In contexts of climate-related mobility, they can be productively understood as part of the politics of adaptation more generally. The politics of adaptation lens provides insight into cross-scalar politics and dynamics around adaptation, as well as the micro-politics level of intra-community relations around adaptation schemes and mobility practices. In our analysis, it helps to highlight how local, informal authorities are integral in terms of shaping adaptation interventions together with authorities at other scales of governance, while prompting contestation from other community groups. This lens breaks the top-down/bottom-up dichotomy often emerging in adaptation literature. Rather, it provides insight into how environmental stress and mobility responses are part of changes in social relations and norms through contestation to particular knowledge, subjectivities or authorities (Eriksen et al, 2015: 525, 530).

Finally, the joint approach offers insight into the differentiated ability of various households and individuals to navigate socioenvironmental change, including adaptation interventions. In the climate literature, this is captured in the concept of differentiated social vulnerabilities (Ribot, 2017; Thomas et al, 2019), and a similar discussion is captured in environmental mobilities literature on 'differentiality of im/mobilities', the 'uneven capabilities and aspirations for im/mobility practices, grounded in both personal and structural factors' (Weigel et al, 2019: 4). In this literature, socioeconomic and demographic characteristics such as generation and gender are often central axes of differentiation that shape climate vulnerability on the one hand (Otto et al, 2017; Rao et al, 2019) and the expectations and legitimacy of certain migration projects on the other (Hunter and David, 2009; Lama et al, 2021). With the latter, such characteristics can inform to what extent staying or leaving is seen as socially acceptable from certain perspectives, as evident in the stances on mobility across generations and destinations evidenced in the preceding quotes. Linking reflections on differentiation in the climate and mobilities literatures helps better situate climate-related mobility within broader socioecological relations. It also offers new analytical and empirical spaces for exploration; for instance, how do differentiated forms of resource access, as highlighted in the adaptation literature, and disparate social and mobility networks, as highlighted in the mobility literature, relate and interact? Also, in terms of outcomes, a combined politics of adaptation/climate mobility perspective draws attention to the far-reaching effects of a 'local' adaptation scheme – that is, how it shapes the flow of people and goods in and out of the affected areas, from local markets to other regions and cities. This is also part of situating adaptation/mobility within larger socio-ecological settings and processes.

Conclusions: Implications for research and policy

This chapter's analysis of irrigation as climate migration management in northern Ghana offers important implications for both research and policy. For research, beyond the utility of the joint politics of adaptation/climate mobility approach described earlier, the case highlights new analytical and empirical foci useful in driving research on climate mobility forward. Foremost is how competing portrayals of climate-related mobility are mobilized and contested in social and policy spheres by diverse actors both across and within scales. Research on climate-related mobility has provided various portrayals of climate-related mobility and those who practise it: climate refugees, climate migrants, climate displaced, trapped populations and so on. Then came critical considerations of these terms, their accuracy and their implications for policy. Now, as particular portrayals of climate mobility are increasingly impacting lived realities, researchers should turn their attention to these portrayals as objects of study – that is, how they are produced, mobilized and contested, by whom and with what effects. This entails linking particular portrayals of climate mobility to concrete sociopolitical contexts and processes, and through these to policy, programming and outcomes on the ground. Such an approach helps to foreground the dynamics of power, politics and representation inherent in how climate-related mobility is portrayed, how particular portrayals gain traction/influence and what tangible outcomes they have when applied in and programming. It also draws attention to the diverse actors involved in the framing and governance of climate-related mobility, from state agencies to various groups within communities themselves.

The chapter's analysis also provides initial policy insights. Our findings indicate that policy efforts to limit climate-related mobility through in situ adaptation programmes will likely be challenging and possibly also detrimental. In the case of the irrigation schemes, mixed and sometimes negative outcomes were linked to limited understanding of the interests, aspirations and influence of different groups within the population, which undermined the ability of projects to address complex and differentiated drivers of mobility. This echoes recent research of the risk of increasing or redistributing vulnerability through adaptation projects (Eriksen et al, 2021) and reinforces the importance of disaggregating 'vulnerable communities', which may be defined by more general indicators of geophysical exposure and poverty. Our analysis also highlights the risk of unintended consequences and maladaptive outcomes directly linked to the irrigation interventions themselves, particularly their poor implementation, management and maintenance.

Together, the preceding findings call for caution in efforts to limit mobility through in situ adaptation projects. Instead, they point to the utility of approaches that support the choice of those in climate affected areas to leave or stay for adaptive outcomes. Supporting choice also entails ensuring representation across diverse and especially highly vulnerable groups that are often underrepresented in the political processes that shape adaptation decisions and practice. Including perspectives from different groups in affected communities will help avoid simplified notions of perceptions and practices of migration as adaptation, supporting greater nuance in policy and programming as these seek to address complex realities of climate change and mobility responses.

Notes

[1] See: https://thediggings.com/gha/map
[2] SADA changed its name to the Northern Development Authority (NDA) in 2017, following the change of government. See https://thecustodianghonline.com/nda-presents-long-term-development-master-plan-to-finance-minister/

References

Abdul-Korah, G.B. (2011) '"Now if you have only sons you are dead': migration, gender and family economy in twentieth century northwestern Ghana', *Journal of Asian and African Studies*, 46(6): 390–403.

Antwi-Agyei, P., Stringer, L.C. and Dougill, A.J. (2014) 'Livelihood adaptations to climate variability: insights from farming households in Ghana', *Regional Environmental Change*, 14: 1615–1626.

Bettini, G. (2017) 'Where next? Climate change, migration, and the (bio) politics of adaptation', *Global Policy*, 8: 33–39.

Bettini, G., Nash, S.L. and Gioli, G. (2017) 'One step forward, two steps back? The fading contours of (in)justice in competing discourses on climate migration', *Geographical Journal*, 183(4): 348–358.

Black, R., Arnell, N.W., Adger, W.N., Thomas, D. and Geddes, A. (2013) 'Migration, immobility and displacement outcomes following extreme events', *Environmental Science & Policy*, 27: S32–S43.

Boafo-Arthur, K. (1999) 'Ghana: Structural adjustment, democratization, and the politics of continuity', *African Studies Review*, 42(2): 41–72.

Boas, I., Kloppenburg, S., van Leeuwen, J. and Lamers, M. (2018) 'Environmental Mobilities: An alternative lens to global environmental governance', *Global Environmental Politics*, 18(4): 107–126.

Boas, I., Wiegel, H., Farbotko, C., Warner, J. and Sheller, M. (2022) 'Climate mobilities: migration, im/mobilities and mobility regimes in a changing climate', *Journal of Ethnic and Migration Studies*, 48(14): 3365–3379.

Bulkeley, H. and Stripple, J. (2013) 'Towards a Critical social science of climate change?', in J. Stripple and H. Bulkeley (eds) *Governing the Climate: New Approaches to Rationality, Power and Politics*, Cambridge: Cambridge University Press, pp 243–260.

Carling, J. (2002) 'Migration in the age of involuntary immobility: theoretical reflections and Cape Verdean experiences', *Journal of Ethnic and Migration Studies*, 28(1): 5–42.

De Haas, H. (2021) 'A theory of migration: the aspirations-capabilities framework', *Comparative Migration Studies*, 9(1): 8.

Eriksen, S.H., Nightingale, A.J. and Eakin, H. (2015) 'Reframing adaptation: the political nature of climate change adaptation', *Global Environmental Change*, 35: 523–533.

Eriksen, S.H. et al (2021) 'Adaptation interventions and their effect on vulnerability in developing countries: help, hindrance or irrelevance?' *World Development*, 141: 105383.

Fagariba, C.J., Song, S. and Soule Baoro, S.K.G. (2018) 'Climate change adaptation strategies and constraints in Northern Ghana: evidence of farmers in Sissala West District', *Sustainability*, 10(5): 1484.

Funder, M., Mweemba, C. and Nyambe, I. (2018) 'The politics of climate change adaptation in development: authority, resource control and state intervention in rural Zambia', *Journal of Development Studies*, 54: 30–46.

Gemenne, F. (2015) 'One good reason to speak of "climate refugees"', *Forced Migration Review*, 49: 70–71.

Gravesen, M., Jarawura, F., Kleist, N., Lindegaard, L.S. and Teye, J.K. (2020) *Governance, Climate Change and Mobility in Ghana*, DIIS Working Paper 2020: 06.

GSS (Ghana Statistical Service) (2018) *Ghana Poverty Mapping Report 'Poverty Trends in Ghana'*, Accra: Ghana Statistical Service.

Hunter, L.M. and David, E. (2009) *Climate Change and Migration: Considering the Gender Dimensions*, Boulder: University of Colorado, Institute of Behavioral Science.

IPCC (Intergovernmental Panel on Climate Change) (2022) *Assessment Report 6, Working Group II: Impacts, Adaptation and Vulnerability*.

Jarawura, F.X., Teye, J., Kleist, N., Lindegaard, L.S. and Quaye, D. (2024) '"These days, things have changed": historicizing current dynamics of climate-related migration in the savannah zone of Ghana', *Climate and Development*: 1–11.

Kleist, N. and Bjarnesen, J. (2019) 'Migration infrastructure in West Africa and beyond', *MIASA Working Paper*, 3: 1–23.

Kleist, N. and Thorsen, D. (eds) (2017) *Hope and Uncertainty in Contemporary African Migration*, New York: Routledge.

Lama, P., Hamza, M. and Wester, M. (2021) 'Gendered dimensions of migration in relation to climate change', *Climate and Development*, 13(4): 326–336.

Lin, W.L., Lindquist, J., Xiang, B. and Yeoh, B.S.A. (2017) 'Migration infrastructures and the production of migrant mobilities', *Mobilities*, 12(2): 167–174.

Lindegaard, L.S. (2018) 'Adaptation as a political arena: interrogating sedentarization as climate change adaptation in Central Vietnam', *Global Environmental Change*, 49: 166–174.

Lindegaard, L.S. and Jarawura, F.X. (2024) *Loss, Damage and Social Cohesion: Impacts and Next Steps for Policy and Programming*, DIIS Report 2024:09. Copenhagen: Danish Institute for International Studies.

Lindegaard, L.S. and Sen, L.T.H. (2022) 'Everyday adaptation, interrupted agency and beyond: examining the interplay between formal and everyday climate change adaptations', *Ecology and Society*, 27(4).

Lindegaard, L.S., Webster, N., Sørensen, N.N. and Borderon, M. (2024) 'Introduction to the special section: unfolding a governance perspective on climate-related mobilities', *Climate and Development*, DOI: 10.1080/17565529.2024.2379464.

Mariama, A. and Ardayfio-Schandorf, E. (2008) 'Gendered poverty, migration and livelihood strategies of female porters in Accra, Ghana', *Norwegian Journal of Geography*, 62(3): 171–179.

Mikulewicz, M. (2018) 'Politicizing vulnerability and adaptation: on the need to democratize local responses to climate impacts in developing countries', *Climate and Development*, 10(1): 18–34.

Mikulewicz, M. (2020) 'The discursive politics of adaptation to climate change', *Annals of the American Association of Geographers*, 110(6): 1807–1830.

Ministry of Food and Agriculture (MoFA) (2018) *Investing for Food and Jobs (IFJ): An Agenda for Transforming Ghana's Agriculture (2018–2021)*.

Nagoda, S. and Nightingale, A.J. (2017) 'Participation and power in climate change adaptation policies: vulnerability in food security programs in Nepal', *World Development*, 100: 85–93.

Nightingale, A.J. (2017) 'Power and politics in climate change adaptation efforts: struggles over authority and recognition in the context of political instability', *Geoforum*, 84: 11–20.

Ojha, H.R., Ghimire, S., Pain, A., Nightingale, A., Khatri, D.B. and Dhungana, H. (2016) 'Policy without politics: technocratic control of climate change adaptation policy making in Nepal', *Climate Policy*, 16: 415–433.

Otto, I.M. et al (2017) 'Social vulnerability to climate change: a review of concepts and evidence', *Regional Environmental Change*, 17: 1651–1662.

Pickbourn, L. (2011) *Migration, Remittances and Intra-household Allocation in Northern Ghana: Does Gender Matter?* Amherst: University of Massachusetts Amherst.

Rao, N., Lawson, E.T., Raditloaneng, W.N., Solomon, D. and Angula, M.N. (2019) 'Gendered vulnerabilities to climate change: insights from the semi-arid regions of Africa and Asia', *Climate and Development*, 11(1): 14–26.

Ribot, J. (2017) 'Cause and response: vulnerability and climate in the Anthropocene', in R. Isakson (ed.) *New Directions in Agrarian Political Economy*, Abingdon: Routledge, pp 27–66.

SADA (Savannah Development Authority of Ghana) (2016) *Infrastructure Development Master Plan*.

Setrana, M.B. and Kleist, N. (2022) 'Gendered dynamics in West African migration', in J.K. Teye (eds) *Migration in West Africa. IMISCOE Regional Reader*, Cham: Springer, pp 57–76.

Schipper, E.L.F. (2020) 'Maladaptation: when adaptation to climate change goes very wrong', *One Earth*, 3(4): 409–414.

Songsore, J. and Denkabe, A. (1995) *Challenging Rural Poverty in Northern Ghana: The Case of the Upper West Region*, Trondheim: Centre for Environment and Development.

Taylor, M. (2015) *The Political Ecology of Climate Change Adaptation: Livelihoods, Agrarian Change and the Conflicts of Development*, New York: Routledge.

Teye, J.K., Awumbila, M. and. Darkwa, A. (2017) 'Gendered dynamics of remitting and remittance use in Northern Ghana', *Migration out of Poverty Working Papers*, 48: 1–36.

Teye, J.K., Jarawura, F., Lindegaard, L.S., Kleist, N., Ladekjær Gravesen, M., Mantey, P. and Quaye, D. (2021) *Climate Mobility: Scoping Study of Two Localities in Ghana*, DIIS Working Paper 2021: 10.

Thomas, K., Hardy, R.D., Lazrus, H., Mendez, M., Orlove, B., Rivera-Collazo, I., Roberts, J.T., Rockman, M., Warner, B.P. and Winthrop, R. (2019) 'Explaining differential vulnerability to climate change: a social science review', *Wiley Interdisciplinary Reviews: Climate Change*, 10(2): e565.

Van der Geest, K. (2011) 'North-south migration in Ghana: what role for the environment?', *International Migration*, 49(s1): e69–e94.

Van Hear, N., Bakewell, O. and Long, K. (2017) 'Push-pull plus: reconsidering the drivers of migration', in N. van Hear, O. Bakewell and K. Long (eds) *Aspiration, Desire and the Drivers of Migration*, Abingdon: Routledge, pp 19–36.

Vandergeest, P. and Peluso, N.L. (1995) 'Territorialization and state power in Thailand', *Theory and Society*, 24: 385–426.

Wiegel, H., Boas, I. and Warner, J. (2019) 'A mobilities perspective on migration in the context of environmental change', *Wiley Interdisciplinary Reviews: Climate Change*, 10(6): e610.

Yaro, J.A. (2004) 'Combating food insecurity in Northern Ghana: food insecurity and rural livelihood strategies in Kajelo, China and Korania', PhD thesis. Oslo: University of Oslo.

8

Short-Distance Migration as an Adaptation Strategy to the Impacts of Climate Change: The Case of the Shashemene District

Zerihun Mohammed

Introduction

As has been established by the earlier chapters in the volume, climate change has become an apparent global reality with profound and multifaceted consequences (IPCC, 2022; WMO, 2022). The problem is more pronounced in developing countries, where people's livelihoods and national economies largely rely on the exploitation of natural resources; however, they often lack the necessary economic, technological and institutional capabilities to devise adaptation and mitigation mechanisms to the adverse impacts of climate change. Even within these countries, the primary victims of the problem are the poor sections of society (Fulco et al, 2007; Malhi et al, 2021).

Ethiopia, like many countries in Sub-Saharan Africa, is among the most vulnerable countries to the impacts of climate change, mainly due to its high levels of dependence on rain-fed agriculture and natural resources for household livelihoods and the national economy. Agriculture provides livelihoods for about 80–85 per cent of the population. It also contributes about 85 per cent of export items and 70 per cent of the raw materials utilized by domestic industries, and above all it employs 80 per cent of the population's labour force (NBE, 2015).

However, this vital sector is significantly affected by the negative impacts of climate change. Long-term climate-change indicators, such as recurrent droughts, the irregularity of rainy seasons and frequent floods, have become common experiences in almost in all regions. Various sources indicate that temperatures have increased by an average of 1°C since 1960. Equally, rainfall showed similar changes in the past two to three decades, with an overall decline trending in almost all parts of the country, with some variations (World Bank, 2021). What is more worrying is that these changes are predicted to worsen in the future and to expose the country to more climate-related environmental, economic, social and even political challenges. For example, the climate risk-assessment profile compiled by the World Bank (2021) indicated that, if things continue unabated, the mean monthly temperature is estimated to rise by 1.8°C by 2050 and by 3.7 °C by the end of the century. Likewise, rainfall is estimated to show a significant decline in all agro-ecological regions of the country, with some variations. Rainfall in the south and centre of the country is projected to show a decrease of up to 20 per cent in three decades' time (World Bank, 2021).

This vulnerability is exacerbated by the relatively low adaptive capacities for dealing with these changes at both the national and household levels (Haileab, 2018; FDRE, 2019; Matewos, 2020). As a result, climate-related phenomena such as recurrent droughts and related severe food insecurity, deforestation, land degradation, and the loss of biodiversity and wildlife will be among the major consequences of climate change in the country (World Bank, 2021).

Impacts of climate change are observable in the Central Great Rift Valley region, where the Shashemene district is located. The impacts of climate change have become more and more apparent in the region (including Shashemene) due to ecological, demographic and economic dynamics. Agriculture, which is the mainstay of most of the region's population, can no longer provide a reliable livelihood base to the ever-increasing population. Meanwhile, it is reported that agricultural productivity is steadily decreasing, while the price of inputs is skyrocketing. In light of the climatic vulnerability and economic uncertainty, farmers are forced to employ various adaptive strategies, including migration, to sustain the wellbeing of their families, particularly to ensure their food security.

This chapter looks at how short-distance rural-urban migration is being used as a strategy of adaptation to the increasing crisis of farming due to climate change. It examines the dynamics of the livelihood base of people in rural Shashemene, the characteristics of those people who go on short-distance migration, its social and economic implications, and why it is preferred to other forms of migration. By looking at these issues, the chapter aims to examine the dynamics and processes of short-distance rural-urban mobility and migration in the research area and to understand the role of climate change in affecting their decision making.

In fact, the links between climate change and migration have attracted the attention of researchers, practitioners and policy makers (Kniveton et al, 2008; Bardsley and Hug, 2010; Martin, 2010; Jayawardhan, 2017; Klepp, 2017). These links are are often complex as they are interwoven with many other intermediate factors, such as household resources, gender composition, age, and access to information. However, there is a consensus that migration, whether domestic or international, is one of the responses to climate change-induced livelihood challenges (Jha et al, 2017; Malhi et al, 2021; Vinke et al, 2022). The key question is how and why people or households exposed to similar types of climate change-related hazards make different decisions in the type of migration.

Research area

Shashemene is located in Oromia Regional State, West Arsi Zone, approximately 250 km south of Addis Ababa. The total land size of the district is 59,586 hectares, of which the rural and urban sections of Shashemene constitute 46,718 hectares and 12,868[1] hectares (Genemo, 2013) respectively. Shashemene is bordered by the districts of Arsi Negele to the north, Shalla and Siraro to the west, Kore to the east, Kofele to the southeast and the newly formed Sidama Regional State to the south. Administratively, the district is divided into two units and is governed by the Municipality of Shashemene (which is responsible for the towns of Shashemene, Kuyera and Bishan Guracha) and the Shashemene Woreda Administration, which administers the rural kebele.

In terms of climatic conditions, the elevation of Shashemene ranges from 1,672 to 2,722 metres above sea level, and its average altitude is around 1,900 metres (Tafere et al, 2006). Local people categorize the district into three loose categories: 'highland' (mainly the northeastern parts of the district bordering Arsi Negele and Kofele), 'lowland' (eastern part of the district neighbouring Siraro and Shalla) and 'midland' (the central kebele in between). While most of the midland and lowlands parts of the district are flat and suitable for agriculture and farming, the highland areas are characterized by slopes of varying degrees. However, the highland area receives more precipitation and produces various types of crops; farmers are said to have relatively better livelihood bases there. The impacts of climate change are more visible in the lowland kebele and have more profound impacts on people's lives. As a result, migration, particularly short-distance migration, is practised more in the lowland than the highland kebele.

The Shashemene district, and its town section in particular, has experienced high population growth in the decades since 1960. The availability of better economic opportunities attracted people to the district from different places, particularly from the highly populated neighbouring districts of Wolyita, Kembata and Hadiya (Benti, 1988). One study found that Shashemene's population growth rate in the 1970s was 9.8 per cent per annum (Bjerén, 1985). Similarly, the 1990s experienced a significant level of demographic increment, with the town experiencing population growth of 6 per cent (Genemo, 2013). The natural increase, coupled with in-migration to the district and the expansion of the town at the expense of the surrounding rural areas, has exacerbated the scarcity of agricultural natural resources such as arable land and pasture available to households.

Research method

Data and information for the study are generated using various data collection methods. The first such method is a survey conducted in three systematically selected rural kebeles, namely Faji Sole, Halache Harabate and O'ine Chefo Umbure. The three kebeles were selected based on their variations in agro-ecology and the resulting possible differences in the intensity of the impacts of climate change. Accordingly, Faji Sole is locally considered to consist of Dega (highland) and has better precipitation and produces different types of crops. O'ine Chefo Umbure is located in the eastern part of the district and is dominated by dry climatic conditions. Precipitation in this part of the district is relatively low and often suffers from shortages of water for both agriculture and domestic use. Halache Harebate is located between the two, close to the town of Shashemene. A total of 410 randomly selected respondents were interviewed for the survey, and data on types of livelihoods, household resources, climate change and its impacts, adaptation strategies, migration and related issues were obtained.

Key informant interviewing was the other data collection method used. Farmers with a history of mobility in the three kebele and migrant labourers in Shashemene and Addis Ababa were identified and interviewed. Accordingly, a total of 15 key informants were interviewed. Four focus group discussions (two male and two female) were also conducted in the rural kebele and the town of Shashemene to obtain collective views of the farms and migrant labourers in the town. Moreover, case studies were used to gain insights into some of the issues relating to the matter in hand.

Climate change and perceptions of climate change in Shashemene

Both metrological data and information from the local people obtained through focus group discussions indicate the presence of slow but steady onset of climate change in Shashemene. An analysis conducted by the Governing Climate Mobility Programme using metrological data for about four decades (1983–2020) indicates that since 1983, the maximum temperature of the district has been increasing by 0.05°C, while the minimum temperature has been decreasing by 0.03°C every year. Likewise, the rainfall pattern showed significant fluctuations in both amount and distribution throughout the months of the year, and the years 1984, 1991, 2003 and 2009 were indicated as years of significantly below-average rainfall.

Farmers in focus group discussions indicated similar changes in temperature and rainfall in the past three decades in all three kebeles. The major indicators they identified as long-term indicators of change are increases in temperature, decreases in rainfall and the unpredictability of the rainy seasons. In all focus group discussions and key informant interviews, farmers indicated a steady increase in temperature in all parts of the district. However, informants in Faji Sole (the highland kebele) emphasized the increase in temperature as more noticeable. They argued that the increase in temperature and reduction of rainfall in the area had forced them to abandon some of the crops they used to cultivate. They cited 'enset' (false banana) as an example and explained that previously the crop had been planted in vast quantities, mainly in homesteads. However, since the 1980s, farmers had been forced to drop it, as it did not grow properly and no longer provided a good harvest. Instead, they increasingly resorted to the cultivation of cereals, such as maize, which used to be a common crop in the midland and lowland parts of the district.

According to the key informants, the problem of climate change is being exacerbated by the increasing population growth and shortage of farmland for young people. Information from the District Agricultural Office indicates that the average farm size of a household in the district is 0.7 hectares per household. As there is no land available to provide to new claimants, gifts from the family have become the only source of land for farming and house making for young people. Moreover, forest, pastureland and marginal lands, such as hillsides, were being converted into farmland to accommodate the increasing demand, which in turn, key informants argued, aggravated the long-term climate-change impacts by causing soil erosion and flooding. High rates of deforestation in state-owned forest patches in Faji Sole kebele and acacia forest in the lowland kebele were also reported.

In conclusion, climate change in Shashemene in general and the three research kebele in particular is evident and clearly affecting people's livelihoods. Informants and focus group discussion participants unanimously agree that the problem has become more serious in the past two decades and even more so in the past five years. As a result, farmers, particularly from poor households, are forced to employ various adaptation strategies, ranging from changes and modifications in their farming system to seeking out other income-earning opportunities both in their own areas and elsewhere so as to sustain the wellbeing of their families, particularly to ensure their food security.

From pastoralism to farming: livelihood dynamics of the Arsi in Shashemene

Traditionally, the livelihoods of the Arsi Oromo in general, and that of the Middle Rift Valley region in particular, were based on cattle rearing (pastoralism). Pastoralism was a better way of life, considering the vast amounts of rangeland at their disposal. Arsi oral traditions indicate that people formerly travelled from place to place seasonally in search of better pasture and water for their cattle, and that farming was less widespread for long periods. With the turn of the 20th century, farming was gradually introduced among the Arsi in the Rift Valley. Although the increasing privatization of rural land and imperial land grants in the area put some of the seasonal rangelands in the hands of other owners and restricted herders' movements, pastoralism still remained the dominant way of life at least until the 1960s. Key informants indicated that, following the expansion of commercial farming in Arsi Negele, Shashemene and Siraro, the Arsi Oromo began to practise farming in addition to cattle rearing.

It was in this situation that the 1975 Land Reform was promulgated, nationalizing all rural lands, including the previous seasonal migration areas of the Arsi; individual farmers were registered in their residence kebele and given farmland with exclusive user rights. As a result, the Arsi began to practise farming regularly, along with the traditional cattle keeping still playing significant role in their livelihoods. With that, Arsi key informants argued, agro-pastoralism (a mixture of farming and cattle-keeping) has gradually become the common way of life since the end of the 1970s.

The shift from pure pastoralism to agro-pastoralism was more rapid in highland areas, where population densities are relatively high, while the relative abundance of land and pasture enables people to keep more cattle in the lowlands. Currently mixed farming has become the main livelihood particularly in the lowland parts of the district and neighbouring woreda. However, the gradual conversion of communal pastureland into individually owned farmland and the absence of new land assigned to pasture have forced people to reduce the number of cattle they keep, making farming progressively more central to the livelihoods of people in all kebeles of the district. Currently, farmers in Shashemene and the neighbouring districts are primarily dependant on farming for their livelihoods.

As explained earlier, the conversion of forest and pasture lands into farmland, the farming of marginal lands and deforestation has increased flooding, the loss of soil fertility, the frequency of crop failure and production losses, as shown in Figures 8.1 and 8.2. As a result, farming in its traditional form is currently facing a major challenge to ensure food security and economic wellbeing for most of the population.

Adaptation to climate change and urban-rural migration as an adaptive measure

In the face of the increasing uncertainties and economic challenges, farmers have used various adaptive strategies, including migration, to cope with the problem. Adaptation to climate change can be defined as actions to reduce

Figure 8.1: Farmland exposed to heavy erosion due to untimely rain in Burra kebele, 25 May 2019

Source: Photo taken by the author

the negative impact of climate change, while taking advantage of potential new opportunities. It constitutes an adjustment in ecological, social and economic systems as a response to the adverse impacts of climate change. It includes changes in practices, processes and structures to alleviate the damage and maximize the opportunities that are available (Burton et al, 2001). It is

Figure 8.2: Failed crop (teff) in Faji Sole kebele due to an absence of rain in the major rainy season, 21 May 2021

Source: Photo taken by the author

a process of adjusting to the actual and expected effects of climate change as a response alongside mitigation. Adaptation can be practised at different levels: country, community and household. It also happens in different shapes and forms, depending on the unique context of a country, community and household.[2]

Besides the climate-change adaptation measures and strategies that countries put in place both nationally and regionally, individual farmers often take actions that would reduce the negative impacts of climate change and ensure the wellbeing of their households. The type, extent and complexity of these adaptive measures vary, depending on the various resources an individual household has at its disposal and the context of local and national social, economic and political conditions. Literature on the area indicates that the types of adaptive strategy an individual farmer employs can be divided into two broad categories, namely 'on-farm' and 'off-farm' adaptation measures (Belay et al, 2017; Bahati et al, 2022; Purwanti, 2022). On-farm adaptations are those implemented with the objective of reducing the risk of production failures and maximizing gain. It includes the use of more and more efficient agricultural practices and inputs, employing soil and water conservation techniques, adjusting farming calendars, shifting cultivation to climate change-resistant crop varieties and moving to the production of high-value crops. The perceived risk level, available resources, levels of knowledge and awareness, and household gender and age composition are among the major factors determining the types and levels of adaptation. Off-farm adaptation measures, on the other hand, are used to address the economic and social challenges of climate change by using different strategies outside the farming system. Engaging in off-farm activities, petty trade and migration are the major types of off-farm adaptation strategies employed by farmers, albeit to different extents (Atube, 2021; Purwanti, 2022; Gemeda et al, 2023).

In fact, migration, whether domestic or international, is a century-old adaptive strategy during crises (González-Hernández et al, 2019; Mwinkom et al, 2021). People move from their places of residence to other places in search of economic opportunities. This can be done on a permanent or temporary basis – that is, for shorter or longer periods of time. Currently, migration in Shashemene has become one of the off-farm adaptation strategies to cope with livelihood challenges created by shortages of land and climate change. However, not all farmers can take migration as an adaptation option. Even those who do take it practise it in different ways (domestic/ international, long-term/short-term, seasonal/permanent and so on) based on the severity of the challenges they individually face, the resources available and the opportunities in their own areas.

Currently, migration practices by farmers in Shashemene can be grouped into three categories based on their destination: international migrants, long-distance rural-urban migrants and short-distance rural-urban migrants. The international migrants from Shashemene are largely women travelling to the Gulf countries to work as domestic workers. In fact, Shashemene and the surrounding Arsi-dominated districts are among the regions with a long history of migration to the Middle East, particularly to Saudi Arabia. The long-distance rural–urban migrants are those who migrate to the major cities, such as Addis Ababa and Adama, in search of jobs. These are largely unmarried young people with very little or no farmland. They often travel to their destinations and stay for relatively longer periods of time (one to three years). The short-distance rural–urban migrants are those who travel to nearby towns, often seasonally, in search of temporary casual jobs.

The focus group discussions with men and women who had migrated from rural to urban areas indicated that short-distance migration largely takes place by commuting from their village of residence to the town of Shashemene and other small towns in the district. This means that most of the migrants make a daily journey from their villages to the jobs market in the towns. A few of them might rent a house as a group and stay in the town on weekdays, going back to their villages on Saturday evening to spend the weekend in their homes. This is practised by both married and single men and women, though for different types of job. In terms of destination, while Shashemene town is the major centre of attraction for the migrants, the small towns of Kuyera and Bishan Guracha (adjacent to Hawassa) also attract some migrants looking for casual jobs.

Short-distance rural–urban migration is preferred to other forms of migration for various reasons. First, it is less costly than international or long-distance migration. In fact, in some cases, it is practised by those who cannot afford to undertake international migration (particularly girls) mainly due to financial constraints and family responsibilities in the case of married women with children. Second, it entails less physical, social and psychological risks than other forms of migration, as those doing so do not travel far from their home areas, culture and social groups.

In addition, unlike long-distance domestic migration, short-distance rural–urban migration has the advantage of maintaining social and economic ties with home villages and allows for frequent back-and-forth movement between the destination and home. This helps migrants to maintain their social and economic ties in their respective villages and allows them to manage to look after the small properties they have.

Almost all male short-distance rural–urban migrants are engaged in casual work in the towns. The construction sector is the primary recruiter of the migrant workers. Some work in petty trade and in loading and unloading trucks and similar activities. There are certain sites in the town of Shashemene where the casual migrant workers assemble early in the morning, and anyone who wants labourers simply goes to these sites and collects as many as he or she wants. The average payment for a day's work in March 2022 fell between 150 and 200 birr (US$2.95–3.95). Although the booming construction sector in Shashemene and the expansion of the town to the rural kebele provided good job opportunities, there is always strong competition for jobs in all areas.

Unlike men, migrant women work in cafés, restaurants and markets. Some of them also engaged in petty trade selling fruit, vegetables, eggs and chickens in small markets. Although there are some who work throughout the year, most of the migrant workers in Shashemene work during the off-farm seasons. The construction sector is the primary recruiter of the migrant workers. As Figures 8.3 and 8.4 show, migrants gather in some pickup places with their own hand tools such as pickaxes and spades to use in the construction jobs available.

Socioeconomic implications of short-distance rural-urban migration

One of the important aspects of rural-urban migration and work as casual labour is the opportunity the migrants have to sustain themselves and their families. Although earnings are low and working conditions are often very harsh, they manage to buy food and other basic needs for the family using the money they earn. Thus, in the absence of any external governmental or nongovernmental support during times of hardship, casual work in towns serves as the only means of survival. This is truer during the years of rain and crop failures. One informant from Faji kebele described the situation as follows:

> There is no rain. We could not farm. Our cattle have died. We lost the mercy of God. To migrate to other areas like others, there is no one looking after our homes and children. We cannot allow us to die. There is no job in this rural area. Thus, our only option is to get up at midnight and walk for three and a half hours to reach Shashemene and try our luck as daily labourers. So, we go out at night and come back at night like wild animals. (Negesso Edao, Faji kebele, Shashemene)

Another informant described the changes in migration from earlier times in the following way:

> Migration is not new to us. We used to live in migration. Our fathers and forefathers used to migrate from place to place with their cattle. By then, they were purely pastoralists. Later, after we learned tilling the land, we used to migrate during the farming seasons. We call that kind of movement 'godanttu' [seasonal migration]. Often, we commute between the highlands and the lowlands with our cattle to our kin. Today, our children are destined for migration. But they migrate to different places and come back with empty hands. This is really bad. This one [going to Shashemene] is a daily wandering, not migration. After all, we spend our nights at home. (Beriso Abdulkadir, Burra kebele, Shashemene)

Figure 8.3: Rural-urban migrants waiting for job opportunities in one of the casual workers' sites, Shashemene, 19 May 2022

Source: Photo taken by the author

The other important implication of the short-distance migration to town is the new trend of Arsi men and particularly Arsi women engaging in casual work. Traditionally the Arsi Oromo were contemptuous of casual working and were very reluctant to work as day labourers. As a result, until recently, most of the casual workers in Shashemene and neighbouring districts were

Figure 8.4: Rural women working as daily labourers cleaning and sorting garlic, Shashemene, 20 May 2022

Source: Photo taken by the author

migrants from adjacent regions (mainly Wolayita, Kembata and Hadiya). As economic hardship exacerbated, the Arsi showed significant attitudinal changes and began to engage in casual work. In fact, this created strong competition between the Arsi (who claim to be local based on the new ethnically based political system) and the old migrant workers, particularly the Wolayita.[3]

More importantly, the engagement of women, both married and young girls, in casual work and petty trade is significant in breaking down the old taboo that expects women to stay at home looking after the children and cattle. In fact, in many cases women have become their families' breadwinners by working in towns. This has brought about a different relationship in families, as some take it as the empowerment of the women.

The engagement of women in casual and petty trade in towns has had negative impacts on children's education, as mothers travel to town and stay there the whole day, while the eldest child, often the daughter, stays at home looking after his or her siblings. This has led to high numbers of school dropouts.

Conclusion

Climate change has significantly affected farming in the study area in recent decades. The problem is reported to have worsened in recent years. Population pressures, a shortage of farmland and other structural problems have aggravated the level of poverty in rural areas and significantly curtailed farmers' coping capacities. In light of the worsening poverty, migration has become an off-farm adaptive strategy.

However, as the preceding case study clearly indicates, the link between climate change and migration is not always direct and straightforward. Households that are exposed to similar types of climate change challenges come up with different responses. These responses are embedded in the economic and social capabilities and constraints of the given household. As a result, migration is a calculated response to climate change by some, while others prefer other forms of responses. Even among those who take migration as a response, the type, duration and length of migration is different.

Domestic migration in the case study is found to be the most readily available option for those who lack the financial resources and/or have local social and economic responsibilities, prohibiting long-distance or international migration. Thus, any policy intervention or practical recommendation in combating climate migration needs to take these intermediate, individual capabilities and constraints into consideration.

Notes

[1] However, this figure is expected to change significantly, since as many as eight rural kebele are in the process of being included in the town and placed under the administration of the municipality.

[2] https://unfccc.int/topics/adaptation-and-resilience/the-big-picture/what-do-adaptation-to-climate-change-and-climate-resilience-mean

[3] One of the causes of the 2017 violent conflict between the Oromo and migrant Wolayita in many parts of the Oromia region is said to be competition for job opportunities. Shashemene was one of the epicentres of the clashes, where a large number of Wolayita were allegedly attacked and looted.

References

Atube, F., Geoffrey, M.M., Martine, N., Daniel, M.O., Simon, P.A. and Okello-Uma, I. (2021) 'Determinants of smallholder farmers' adaptation strategies to the effects of climate change: evidence from northern Uganda', *Agriculture and Food Security*, 10(6): 1–14.

Bahati, M.A., Mohan, G., Matsuda, H., Melts, I., Kefi, M. and Fukushi, K. (2022) 'Understanding the farmers' choices and adoption of adaptation strategies and plans to climate change impact in Africa: a systematic review', *Climate Services*, 30: 100450.

Bardsley, K.D. and Hug, G.J. (2010) 'Migration and climate change: examining thresholds of change to guide effective adaptation decision-making', *Population and Environment*, 32(2): 238–262.

Belay, A., John, W.R., Teshale, W. and John, F.M. (2017) 'Smallholder farmers' adaptation to climate change and determinants of their adaptation decisions in the Central Rift Valley of Ethiopia', *Agriculture & Food Security*, 6(24): 1–13.

Benti, G. (1988) 'A history of Shashemene from its foundation to 1974', MA Thesis. Addis Ababa: Department of History, Addis Ababa University.

Bjerén, G. (1985) *Migration to Shashemene: Ethnicity, Gender and Occupation in Urban Ethiopia*. Uppsala: Scandinavian Institute of African Studies.

Bjerén, G. and Atakilte, B. (2009) 'Change and transformation in Shashemene town 1973–2008'. Unpublished conference paper, 16th International Conference of Ethiopian Studies, Addis Ababa University.

Burton, I. et al (2001) 'Adaptation to climate change in the context of sustainable development and equity', in J.J. McCarthy et al (eds) *Climate Change 2001: Impacts, Adaptation and Vulnerability*. Cambridge: Cambridge University Press, pp 879–912.

Federal Democratic Republic of Ethiopia (FDRE) (2019) 'Ethiopia's Climate Resilient Green Economy National Adaptation Plan'. Addis Ababa: Federal Democratic Republic of Ethiopia.

Fulco, L., van Scheltinga, C.T., Verhagen, J. and Kruijt, B. (2007) *Climate Change Impacts on Developing Countries: EU Accountability European Parliament's Committee on the Environment*. Wageningen: Wageningen University and Research Centre.

Gemeda, O.D., Diriba, K. and Weyessa, G. (2023) 'Determinants of climate change adaptation strategies and existing barriers in southwestern parts of Ethiopia', *Climate Services*, 30: 100376.

Genemo, B. (2013) *Remote Sensing and GIS Based Urban Sprawl Susceptibility Analysis: A Case Study of Shashemene Town, West Arsi Zone, Ethiopia*. LAP LAMBERT. Saarbrücken: Academic Publishing.

González-Hernández, D.L., Meijles, E.W. and Vanclay, F. (2019) 'Factors that influence climate change mitigation and adaptation action: a household study in the Nuevo Leon region, Mexico', *Climate*, 7(6): 76.

Haileab, Z. (2018) 'Climate change in Ethiopia: impacts, mitigation and adaptation', *International Journal of Research in Environmental Studies*, 5(1): 18–35.

IPCC (Intergovernmental Panel on Climate Change) (2022) *Climate Change 2022: Impacts, Adaptation and Vulnerability – Technical Summary*. Cambridge: Cambridge University Press.

Jayawardhan, S. (2017) 'Vulnerability and climate change-induced human displacement', *Consilience*, 17: 103–142.

Jha, C.K., Gupta, V., Chattopadhyay, U. and Sreeraman, B.A. (2017) 'Migration as adaptation strategy to cope with climate change: a study of farmers' migration in rural India', *International Journal of Climate Change Strategies and Management*, 10(1): 121–141.

Klepp, S. (2017) 'Climate change, drought, land degradation and migration: exploring the linkages', *Oxford Research Encyclopaedia, Climate Science*. Oxford: Oxford University Press.

Kniveton, D., Schmidt-Verkerk, K., Smith, C. and Black, R. (2008) *Climate Change and Migration: Improving Methodologies to Estimate Flows*. Brighton: University of Sussex.

Malhi, G.S., Kaur, M. and Kaushik, P. (2021) 'Impact of climate change on agriculture and its mitigation strategies: a review', *Sustainability*, 13(3): 1318.

Martin, S. (2010) 'Climate change, migration, and governance', *Global Governance*, 16(3): 397–414.

Matewos, T. (2020) 'State of local adaptive capacity to climate change in drought prone districts of rural Sidama, southern Ethiopia', *Climate Risk Management*, 27: 100209.

Mwinkom, X.K.F., Damnyag, L., Abugre, S. and Alhassan, S.I. (2021) 'Factors influencing climate change adaptation strategies in North-western Ghana: evidence of farmers in the Black Volta Basin in Upper West region', *SN Applied Sciences*, 3(548).

NBE (National Bank of Ethiopia) (2015) *National Bank of Ethiopia 2014/2015 Annual Report*. Addis Ababa: NBE.

Purwanti, T.S., Syafrial, S., Huang, W.C. and Saeri, M. (2022) 'What drives climate change adaptation practices in smallholder farmers? Evidence from potato farmers in Indonesia', *Atmosphere*, 13(1): 113.

Tafere, Y., Tadele, F. and Lavers, T. (2006) *Arada Area (Kebele 08/09)*. Shashemene: Ethiopian Urban Studies.

Vinke, K. et al (2022) 'Is migration an effective adaptation to climate-related agricultural distress in Sub-Saharan Africa?', *Population and Environment*, 43: 319–345.

WMO (World Metrological Organization) (2022) *State of the Global Climate 2022*, Report Number WMO-No. 1316.

World Bank (2021) *Climate Risk Country Profile: Ethiopia*. Washington DC: World Bank Group.

PART III

Gendered Mobility Practices

9

Gender Dimensions of Climate Change-Related Migration in the Savannah and Forest Agro-Ecological Zones in Ghana

Francis Xavier Jarawura, Nauja Kleist and Joseph Kofi Teye

Introduction

While it is increasingly recognized that climate change may result in significant changes in global migration patterns due to its aggravating effects on the security of human populations, this is expected to be more prominent in the developing world, where livelihoods are still largely based on or related directly to ecosystem resources, with little or no irrigation available for growing crops and rearing animals (IPCC, 2014; Warner and Afifi, 2014; Riede et al, 2016). Developing countries also have high levels of poverty that limit their ability to withstand and adapt their livelihoods to changing climate circumstances (Shackleton et al, 2015; Lawson et al, 2019). The Intergovernmental Panel on Climate Change (IPCC) (2018) emphatically states that climate change is already influencing human migration across the globe. This is largely due to the wide range of drivers of migration and the complex nature of their interaction with environmental factors (Obokata et al, 2014; Hummel, 2016; Beine and Jeusette, 2021). Comprehensive empirical evidence on the subject is nevertheless still limited.

Climate events and processes will rarely act independently in the migration process. Rather, they will often act in concert with economic, social and political drivers, which themselves affect migration. This situation will make it very difficult to single out migrants for whom climate change is the primary driver ('environmental migrants'), a trigger or a background factor ('climate-related migrants') (Foresight, 2011; McLeman, 2013; Kaczan and

Orgill-Meyer, 2020). While studies of the general links between climate change and migration are growing, the gender dimensions have seen little attention. This is against the background that climate change could have aggravating consequences for human population mobility with differentiated gendered effects (Foresight, 2011; IPCC, 2014). Gendered processes and conditions in societies are crucial in shaping vulnerabilities and mediating climate impacts and responses such as migration. Although there is growing (albeit limited) attention to the gender dimensions of climate change and migration (Wrigley-Asante et al, 2019), there are few comparative studies that help us understand how a different context might influence the nexus.

This chapter presents a comparative analysis of gender and climate mobility between two spatially and culturally differentiated regions: the Upper West and the Eastern regions in Ghana. The former region is located in the Forest zone, while the latter is located in the Savannah zone. The main cultural difference between the two is that ethnic groups in the Savannah zone have a patrilineal inheritance system, while those in the Forest zone have both patrilineal and matrilineal inheritance. Studies on climate change and (gender) migration in Ghana have largely concentrated on the north of the country due to the high-risk nature of the climate (Dickson and Benneh, 1988, Dietz et al, 2004), the relatively high and persistent poverty (Ghana Statistical Service (GSS), 2018) and the high levels of out-migration from the zone compared to the rest of the country (van der Geest, 2011). However, there is growing evidence of unprecedented climate change in the Forest zone with concomitant emerging patterns of migration.

The maximum mean temperature in the Forest zone has risen on average by 0.027°C annually compared to 0.023°C in the Savannah zone between 1970 and 2018 (Teye et al, 2021). The authors also observed changes in the migration patterns in the Forest zone associated with climate change. They claim that changes in temperature and rainfall have rendered crop farming less lucrative, with local people responding in many ways, including migrating out in search of better climates and nonfarm jobs.

While the higher rate of temperature change in the Forest zone and its association with migration does not necessarily suggest greater vulnerability compared to the Savannah, it does call for increased attention to understand its impacts and the responses of those affected. Furthermore, besides the observed physical differences between the Savannah and Forest zones, cultural differences also abound. Thus, comparing the two extreme environments, in terms of how the inhabitants experience and respond to climate change, can help discern differences and commonalities as well as synergies that can feed into both local and national planning.

The chapter is organized as follows. The next section establishes the relations between climate change, gender and migration. This is followed by a section presenting the research areas. The next section outlines the methods of enquiry used for the chapter and is followed by a section presenting the findings. The final section presents the study's major conclusions.

Climate change, gender and migration

Although climate change is physical in nature, its effects impact the social, economic and political realms of life. The physical dimensions of climate change are mediated by socioeconomic conditions in determining vulnerability outcomes (Adger et al, 2007; Mertz et al, 2009). Vulnerability to climate change is thus highly contingent on socioeconomic conditions (Bohle et al, 1994). Climate change vulnerability here refers to 'the degree, to which a system is susceptible to, or unable to cope with adverse impacts of climate change, including climate variability and extremes' (IPCC, 2007: 883). As socioeconomic conditions vary for different social categories of people, vulnerability to climate change differs among and within different social groups such as men and women. Recognition of the effects of climate change as being shaped by gendered realities (Djoudi and Brockhaus, 2011) is crucial for understanding how men and women in different localities experience and respond to climate change effects (IPCC, 2018).

Globally, women are often more vulnerable to climate change, yet have limited capacity to respond. This is due to differential access to resources emanating from social processes such as inheritance rules, tenure systems and gender roles (Bawakyillenuo et al, 2016; Wrigley-Asante et al, 2019). Recognizing and understanding differential vulnerability to climate change among different groups in society is a crucial step in dealing with the unfolding climate crisis. The reconfiguration of power relations is crucial in mediating both vulnerabilities and capacities, and the preferences of different groups, such as men and women (Christiansen et al, 2007). The contours of power relations are critical in shaping experiences, perceptions and adaptive capacities and the flow of external assistance for both men and women (Agrawal et al, 2008). Moreover, gender reproduces vulnerabilities that mediate these outcomes (MacGregor, 2010).

One coping or adaptation strategy that is critical to consider between men and women is migration (Foresight, 2011; IPCC, 2018). While many studies have attempted to establish the relationship between climate change and migration, few have given attention to the gender dimensions of the phenomenon (Lama et al, 2021). Unequal gender relations and access to resources may render women more vulnerable to climate change than men. Given the disadvantaged position of women in society as people with relatively limited access to resources, lower incomes, less education and less decision-making power, they will be disproportionately affected by climate change (IPCC, 2007; Demetriades and Esplen, 2010). Although this emphasis on women as a vulnerable group has the potential to highlight the magnitude and potential impacts of climate-related gender migration, there is a greater risk of ignoring the role of 'the root causes of migration and [the] everyday practices that shape unequal vulnerabilities over time' (Lama et al, 2021: 327). This helps to draw attention to the need to see gender constantly as an important factor amid a myriad of drivers of migration. This allows for a more focused inquiry into migration as a process that is constantly influenced by dynamic processes of change. Lama et al (2021) also argue that the physical movement itself, as in temporary to permanent, seasonal to permanent, from voluntary to forced, has received more attention than the 'various social meanings that mobility represents and reproduce, such as freedom, justice and opportunities' (Lama et al, 2021: 326). It is thus crucial to understand how various social processes become intertwined in shaping vulnerabilities and how climate change effects are socially mediated and influence mobility patterns, and what these patterns represent.

In Ghana, the increasing climatic variability, spatial configuration of poverty and internal gender relations are crucial in shaping climate change vulnerabilities and adaptations across different regions and among social groups (Yaro et al, 2015). Spatially, the effects of climate change are more critical in the Savannah, where more people depend on rain-fed farming for a livelihood. The area is also the most deprived in terms of physical and socioeconomic infrastructure. Although the Forest zones are relatively wealthier, they still lack the conditions necessary for effective and sustainable climate adaptation (Teye et al, 2021). These structural constraints are funnelled into local cultural gender relations and differences shaped by traditional norms. Cultural norms define the configurations of vulnerability to climate change in many ways. In the Savannah, patrilineal inheritance and therefore male dominance ensures that men have more resources and are therefore more able to respond to the effects of climate change in better and more diverse ways than women (Bawakyillenuo et al, 2016). In Ghana's Forest zone, both patrilineal and matrilineal inheritance exist. This is expected to produce a mixed outcome in terms of how both men and women experience and adapt to climate change. Also, culturally determined ways of farming in both the Savannah and Forest zones limits the crops and animals that women can farm. This has consequences for the range and type of adaptation practices that women specifically can adopt. In terms of migration as a response to climate change, differential experiences, perceptions and resources, gender roles, and mobility in society could influence the participation of males and females differently.

Study areas

Primary data were collected from Wa West, a district in the Upper West region, and Yilo Krobo District of the Eastern region. Two communities were selected from each of the two districts (see Figure 2.1 in Chapter 2). Wa West has a typical Savannah geography with a generally flat topography and very few highlands. The district is mainly drained by the Black Volta River and its tributaries. Like the rest of the Savannah, the district has sparse vegetation and a dry climate. Common trees found in this region include shea, dawadawa, neem and baobab, which are all drought-resistant. The district experiences a single rainy season, which occurs from April to September, with an average annual rainfall of about 115 cm. A single rainy season concentrates the risks of rain-fed farming to just one season. About nine out of every ten people (86 per cent) are engaged in farming, which is almost entirely rain-fed. The common crops grown include maize, groundnuts, sorghum, rice, bambara beans and soya beans. Livestock that are reared include goats, sheep, pigs, cattle and poultry (GSS, 2012). Besides farming, the people also engage in mostly seasonal, circular and permanent forms of migration, mostly to the south of the country. Indeed, the region has a long history of migration largely linked to colonial development policies and, to an extent, its environment (van der Geest, 2011).

Yilo Krobo district has a diverse topography, from low-lying areas to highlands. The district is mainly drained by Lake Volta, which covers large stretches of the land. The district is classified under the Forest agro-ecological zone, which is characterized by forest vegetation and a tropical climate. The vegetation is mainly made up of coconut and palm trees. The climate is characterized by a bimodal rainfall pattern from March to October, and again from November to February, with a mean annual rainfall of around 1,800 mm (Ofori-Sarpong and Annor, 2001; GSS, 2012). The bimodal rainfall pattern in the district makes farming less risky compared to farming in the Wa West district. The main crop grown is cocoa. Climate changes, land-use and cover changes and soil degradation are progressively causing a change from cocoa cultivation to food-crop farming (Attua, 2003). Other crops cultivated include pineapple, banana, mango, maize and cocoyam.

Methodology

This chapter draws on quantitative and qualitative data collected from two districts in the Upper West region and the Eastern region. We conducted a survey of 469 individuals (243 males and 226 females) from six villages based on the proportions of their total disaggregated populations of males and females. Using a semi-structured questionnaire, we collected data on the nature of economic strategies, governance and relations, perceptions of climate adaptation to climate change, and migration. We employed focus group discussions (FGDs) and individual in-depth interviews to collect the qualitative data. In each village we conducted two FGDs with men and women respectively. FGD participants consisted of village subchiefs, leaders of associations and ordinary people, in all numbering between 11 and 16 people. By means of the FGDs, we gathered information on general village wellbeing, livelihood strategies, stressors both internal and external in nature, changes in climate, and the responses of households and institutions.

We also conducted in-depth interviews in each of the villages to understand how household assets and decisions are shaped by both traditional and modern institutions. Institutions shape the context in which households draw resources and make decisions for adaptation (Agrawal, 2008). In every village, we interviewed four male and four female household heads. This approach is essential for gaining diverse perspectives and realities from a gender perspective. We also interviewed two subchiefs in each village to understand how local structures and norms influence the adaptive capacities of different households.

Perceptions of climate change

Perceptions of climate change are critical to how people respond to environmental phenomena (Yaro et al, 2015; Popoola et al, 2018). Therefore, studying perceptions of key climate variables, including temperature and rainfall, can help us understand the actions and intentions of rural actors. As temperature and rainfall are the key climate variables influencing agrarian livelihoods in West Africa (Owusu and Waylen, 2009; Teye et al, 2015), this study delved into respondents' perceptions of them, as this helps us understand their responses better. As shown in Table 9.1, respondents, both male and female, in both regions perceived temperatures as generally having risen over the last 20–30 years. However, respondents in the Upper West region, both male and female, perceived temperatures as increasing nearly 13 per cent more than those of the Eastern region's respondents. These perceptions reflect the general landscape of research findings in both regions. For example, in their study of climate change in the two regions, Teye et al (2021) conclude that both the minimum and maximum temperatures have generally risen since 1970. However, they also note that the warming trend was increasing faster in the Eastern region. The annual mean maximum temperature in the region has been rising at 0.027°C compared to 0.023°C in the Upper West region. The rapid change in the Eastern region is having significant effects on the cocoa sector, as the crop experiences lower output under higher temperatures, with consequences for migration (Teye et al, 2021).

In terms of rainfall, respondents from both regions, both male and female, generally perceived a declining trend in rainfall over the last 20–30 years. However, vast differences exist between them in terms of the extent of the decline in rainfall. As shown in Table 9.2, while about eight out of ten respondents in the Upper West region considered rainfall to have declined, only about five out of ten in the Eastern region did so. The proportion of males and

Table 9.1: Perceptions of temperature

| | Yilo Krobo | | | | | | Wa West | | | | | | |
| | Male | | Female | | Total | | Male | | Female | | Total | |
	N	%	N	%	N	%	N	%	N	%	N	%
Increased	76	62.8	80	73.4	156	67.8	102	83.6	90	76.9	192	80.3
Decreased	21	17.4	12	11	33	14.3	14	11.5	21	17.9	35	14.6
No change	17	14	9	8.3	26	11.3	5	4.1	3	2.6	8	3.3
Don't know	7	5.8	8	7.3	15	6.5	1	0.8	3	2.6	4	1.7
Total	121	100	109	100	230	100	122	100	117	100	239	100

Source: Governing Climate Mobility Household Survey data, 2020

Table 9.2: Perceptions of rainfall

	Yilo Krobo						Wa West					
	Male		Female		Total		Male		Female		Total	
	N	%	N	%	N	%	N	%	N	%	N	%
Increased	43	35.5	30	27.5	73	31.7	22	18	23	19.7	45	18.8
Decreased	58	47.9	51	46.8	109	47.4	98	80.3	90	76.9	188	78.7
No change	13	10.7	17	15.6	30	13	1	0.8	2	1.7	3	1.3
Don't know	7	5.8	11	10.1	18	7.8	1	0.8	2	1.7	3	1.3
Total	121	100	109	100	230	100	122	100	117	100	239	100

Source: Governing Climate Mobility Household Survey data, 2020

females who perceived declining rainfall is higher in the Upper West region than the Eastern region. Their perceptions reflect the findings of many scholars that point to higher levels of climate variability and vulnerabilities in the Upper West region (van der Geest, 2011). Only about a third of respondents considered rainfall to be rising over the period. These varied responses are not surprising, as people may experience differential rainfall effects because they inhabit different microclimates and soil zones and also grow different types of crops.

Generally, the perceptions of the respondents are consistent with the general literature, which points to a declining trend in rainfall in Ghana (Yaro et al, 2015), which could contribute to changes in migration patterns (Teye and Owusu, 2015). What appears to be more crucial for respondents regarding rainfall is not the amounts, but the distribution. The consensus in both regions is that rainfall had become more variable and much more difficult to predict. Focus groups did not lack clarity in expressing the nature of the distribution, as noted in the following statement:

> So, over the last 20 years it has been raining better than when I was young, but the problem is that it rains more heavily these days, with more floods, and it does not come regularly too. But this is still better than the years of the big droughts [the 1980s]. Now we say we get more rain, but there are also a lot of floods and small droughts that worry us a lot. (Elderly man, Upper West FGD)

This statement reflected the collective fact that respondents have not been oblivious to the fact that rainfall amounts have been increasing from the 1990s amid high intra-annual variations experienced in terms of weather extremes such as floods and frequent though less intense dry spells and frequent but shorter and less intense droughts, as noted by several reports, including Teye et al (2021) and the IPCC (2018).

Gender dimensions of climate-related mobility

Climate change as a driver of migration

Discerning the gender dimensions of climate-related migration requires an exploration of existing patterns of migration and the reasons or drivers explaining the movements of both men and women, boys and girls (Foresight, 2011). This provides a broader entry point from which we can then delve into various specifics of gender and climate-related migration. Thus, we explored the various drivers of migration among men and women. We also explored the adaptation strategies employed by each group to help reveal the links to migration. Focus groups, interviews and survey results point to different but complex links between climate adaptation and migration, the degree of involvement of different genders, and other associated drivers.

Table 9.3: Gender and climate-related forms of migration in the Wa West district (Upper West region), frequency (%)

Type of migration	Male			Female		
	Yes	No	N/A	Yes	No	N/A
Seasonal migration to urban areas	46(37.7)	76(62.3)	0(0)	20(17.1)	95(81.2)	2(1.7)
Migrated sometimes to urban areas	32(26.2)	86(70.5)	4(3.3)	15(12.8)	98(83.8)	4(3.4)
Seasonal migration to rural areas	32(26.2)	88(72.1)	2(1.6)	22(18.8)	90(76.9)	5(4.3)
Migrated sometimes to rural areas	36(29.5)	78(63.9)	8(6.6)	29(24.8)	82(70.1)	6(5.1)
Short- or long-term migration abroad	24(19.7)	92(75.4)	6(4.9)	4(3.4)	103(88)	10(8.5)

Source: Governing Climate Mobility Household Survey data, 2020

From a generic point of view, the multiplicity of the drivers of migration is well acknowledged by respondents in both the Upper West region and the Eastern region. These drivers include: poverty, marriage, a poor harvest, unreliable rainfall, the growing difficulty of predicting rainfall, the growing unreliability of 'rain-callers', increasing dry spells and droughts, soil impoverishment, escaping witches and bad 'juju' (magic), escaping family feuds, the quest to see the city, the desire for higher incomes, and interest in other livelihood opportunities (largely due to education or new skills) that are non-existent in their original homes. The identification of climate drivers is unsurprising, as both regions depend primarily on rain-fed agriculture for their livelihoods (Teye et al, 2021). Moreover, poverty as a driver of migration also encompasses the role of climate as depicted by focus groups and interviews. However, the explanation varies among the respondents, as some refer to absolute poverty situations, while others use it to mean relative poverty or inequality. Absolute poverty is largely a function of crop productivity, which relies heavily on climate. Yet, inequalities abound among households, as they have varied capacities and a variety of secondary sources of income. As will be shown later on, inequality is viewed as an important driver of migration, especially for the young, as poorer individuals try to migrate to earn an income to match those who are seen to be wealthier. Generally, focus groups and interviews identify rural-urban migration as the dominant stream, but still highlight the growing importance of rural-rural migration. The former is closely linked to what focus groups generally term the 'modernization opportunity', where young people are attracted to urban and city locations to earn nonfarm incomes. Focus groups emphasize that,

Table 9.4: Gender and climate-related forms of migration in the Yilo Krobo district (Eastern region), frequency (%)

Type of migration	Male			Female		
	Yes	No	N/A	Yes	No	N/A
Seasonal migration to urban areas	16(13.2)	101(83.5)	4(3.3)	4(3.7)	96(88.1)	9(8.3)
Migrated sometimes to urban areas	36(29.8)	80(66.1)	5(4.1)	16(14.7)	86(78.9)	7(6.4)
Seasonal migration to rural areas	12(9.9)	100(82.6)	9(7.4)	8(7.3)	94(86.2)	7(6.4)
Migrated sometimes to rural areas	14(11.6)	98(81.0)	9(7.4)	14(12.8)	88(80.7)	7(6.4)
Short- or long-term migration abroad	2(1.7)	114(94.2)	5(4.1)	1(0.9)	105(96.3)	3(2.8)

Source: Governing Climate Mobility Household Survey data, 2020

amid the challenges, including climate change, in the agrarian sector, the young see better opportunities to make much more money in urban areas more quickly, as noted in the following statement:

> These days the rainfall has been changing and bringing more short spells and making it difficult to farm, so young and middle-aged women and men (and girls and boys), who want money do migrate to towns to work. And they do that even more when there is low income here from farms during poor rainfall. (Female, FGD, Upper West region)

It is explained that, besides the job opportunities provided by urban places, the allure of their beauty, the available social amenities and the modern lifestyle such as freedom of association, the variety of foreign foods and income autonomy is crucial in how young people make decisions. It is also felt that women and girls in particular find migrating liberating from longstanding social norms that stand in their way when they want to take important decisions for themselves. This highlights the importance of intervening variables in the decision to migrate when faced by negative climate circumstances. Other works, such as Jarawura and Smith (2015), have highlighted the role of the attractiveness of urban infrastructure and lifestyle in the climate-migration nexus. In difficult or desperate moments, rural dwellers do not just go where they can find jobs or income and food, but also where they can meet other wants and needs such as leisure, social freedom and economic autonomy. A generic exploration of the drivers of migration already shows that connections exist between climate change and migration. However, these connections are far from being simplistic; rather, they are complexly intertwined with other social processes. Unravelling the complexity or at least illuminating the context in which migration decisions are taken under changing climatic circumstances thus requires a more holistic approach on the adaptation measures employed by respondents.

Respondents were asked to indicate the different means by which they tried to adapt to climate change effects. Migration turned out to be an important strategy with spatial variations. The majority of respondents who had engaged in all types of climate-related migration were from the Upper West region (see Tables 9.3 and 9.4). Also, a greater proportion of males than females had engaged in all types of climate-related migration in the two regions. A considerable proportion of males (37.7 per cent) from the Upper West region had engaged in seasonal (annual) migration to urban areas in comparison to females (17.1 per cent), while few males (13.2 per cent) and females (3.7 per cent) from the Eastern region had engaged even once in a similar type of migration pattern. Compared to respondents (both male and female) from the Upper West region, more males (29.8 per cent) and females (12.8 per cent) from the Eastern region sometimes migrated to urban areas. Also, more respondents (male and female) from the Upper West region engaged in seasonal migration to rural areas, as well as sometimes migrating to rural areas more than respondents in the Eastern region.

The survey results highlight the role of climate in driving different forms of migration and their gender dynamics. This is also reflected and in much detail in the accounts of our focus groups and interviews. In both regions, respondents identify climate change as an important driver of changes in migration patterns. They point to both the intensification and diversification of migration patterns as climate change intensifies the frequency of dry spells, drought and floods, as noted in the following statements:

> These days, the rains are difficult to understand, unlike the 1990s to the 2010s. It rains a lot now, but the problem is that it stops more often, and by the time it comes back the crops have already wilted. Also, there are increasing short spells and floods causing us poor harvests ... Those days our older men used to migrate to solve these situations, but now even young people and women, as well as small boys and girls, have joined the migration ... we the women too are affected because if there is nothing to harvest then we have no jobs and no money. Even the man struggles to care for you, so you must also wake up [do something unusual]. (Elderly woman, FGD, Upper West region)

> Over here, it used to be only educated people migrating to the cities and also those young men who wanted to see the city and come back after some time. Those days, when there was poor rainfall, our

brothers used to go migrate to the Western region to work and bring back food. The women were not part of this, but today you can't stop them. (Elderly woman, FGD, Eastern region)

But these days the rains are much less and irregular than 20 years ago. There are more problems with increasing occurrences of drought, and the soil too is losing its fertility, so if there is not enough rain, nothing grows well. If we don't harvest well, we the men lose our work and money and can't cater for the women and children, and the women too lose the small jobs that were available, so that is why people are migrating, and they don't only wait for bad rains, they go because it can happen. The younger ones want to go to the city and do intelligent things, but some of our people, especially the old ones, know nothing but our farming, so we go and look for land elsewhere, with good rain and soil. (Woman, FGD, Eastern region)

These statements highlight some of the imperatives emphasized by focus groups and interviews. First, they clearly show the effects of climate on livelihoods and how both men and women (and boys and girls) are influenced to migrate based on cultural norms. In the Upper West region, patriarchal norms make it the practice for men to inherit household or family properties, to head households, control resources and make major decisions. This means that in the event of their own production failure, the onus falls on them to salvage the situation in other ways such as migration. This situation is also played out in the same manner in the study areas in the Eastern region, even though there is mixed patrilineal and matrilineal inheritance in different clans. Among some clans of the Eastern region, inheritance is not by the male line, but rather through the female line. This acts as a disincentive for some women and girls to migrate when confronted with crises such as those posed by climate change. There is a trend towards modernization that allows women and girls to act in the migration streams, with or without the consent of the family head, especially when the head fails in his obligations towards them or when they lose their own income-earning opportunities. The literature on migration has noted this trend clearly. For example, Adepoju (2010) and Zhao et al (2023) have noted the increasing feminization of West African migrations and have attributed it partly to education and the breakdown of social norms that inhibited women and girls in particular from migrating. The nature of the gender norms surrounding migration are constantly being transformed, as reflected in the words of one young seasonal migrant:

They say that we have to stay home to fetch water, cook and wash bowls and hold crying babies, while our brothers are out there in Accra making money for their education and other needs. When there was so much flooding about ten years ago, myself and my auntie just sneaked out and went to Kumasi … to join our brothers who had already left … and we also brought something home for ourselves and our mother, and they can't do anything to you like the old days. (A 28 year-old woman, interview, Upper West region)

This extract highlights social change in the society where traditional norms are breaking down and there is growing female interest in migration. This and other statements also highlight how migration responses to climate crises have been transformed from a male-dominated enterprise to one involving both genders, and it draws attention to other intervening factors, such as soil infertility, in the decision to migrate. Furthermore, the statements show that in relation to climate change, migration is driven not only by real but also by perceived effects on crops and livelihoods. This alludes to the fact that livelihood adaptations, including migration, are in part a function of anticipation based on previous experiences (Scoones, 1998). This has only recently been acknowledged by the climate-migration literature (Jarawura and Smith, 2015). The real but also anticipated effects of climate change on crop production in both regions have been at the epicentre of climate-related migration.

In the Upper West region, where there is a single maxima rainfall regime with negligible irrigation (GSS, 2018), the farming season is at the mercy of rainfall. The region has experienced frequent droughts, flooding and dry spells over the last four decades, with serious effects on food security and migration. As men and women grow different crops, the effects of climate change and their timing also differ. Men mostly grow staples, including maize, millet, cassava, beans and groundnuts, while women mostly grow vegetables and groundnuts. Accounts of focus groups indicate that the main women's crop, groundnuts, is much more susceptible to dry spells and drought when it is flowering compared to those of men. Thus, accordingly, women begin to plan their migration much earlier than men when there is poor rainfall. As for the men, 'they only do this when there is total wilting or poor fruit, but you see it can happen that women lose all their groundnuts, but the men will never easily lose all because they cultivate many crops that respond differently to poor rains', explained an elderly woman from the Upper West region. This situation also occurs in the Eastern region. While restrictions on women growing men's crops have loosened over time, many obstacles, such as their household workload and lack of or poor capital, still stand in their way. Women and girls also depend on the success of men's crops in order to be catered for by the men and to obtain farm employment with payment in kind or in cash. Thus, women and girls are confronted with migration decisions with either male or female crop failures. This highlights the disproportionate impacts of climate change on females and how their migration in relation to climate change is shaped differently.

In the Eastern region, the role of climate change as a driver of migration largely lies in effects on the cocoa, mango and vegetable sectors. First, the declining rainfall and rising temperatures are said to be negatively affecting production generally. Second, declining rainfall affects cocoa farmers far more than those in households that depend largely on mango and food crops, thereby altering both absolute and relative poverty. This has two effects: first, it has contributed to a declining trend in income from cocoa so that farmers migrate to the Western region (a neighbouring administrative region) to establish farms, as the area has a better microclimate; and, second, young people in cocoa and food crop-farming households are becoming less wealthy than those in mango-farming households, prompting the former to migrate in search of more income in response to the changing configurations of wealth. Mango and food crop farming are largely new strategies developed in response to a range of factors, not least the changing climate.

Rural-urban migration

In conformity with the survey results, focus groups and interviews in both regions note a rising dominance of rural–urban movements championed by men. One driver of this process is the decline of rural job opportunities for males at traditional destinations in southern Ghana due to the increasing use of agro-chemicals in controlling weeds, which reduces the need for farm hands. This is clearly noted by one of our respondents:

> This time, when there are droughts and floods, or the rains are late, the women who migrate find more job opportunities than the men in the rural areas. The use of herbicides and weedicides for farming has scrapped the opportunities that the men used to get when they travel. So, when they travel down to the rural south, they hardly get work to do. But for the women, they can do sowing, harvesting, cooking on the farm, cashew-nut collection and many others works ... and men cannot do these works ... how can a man cook or pick cashew nuts? (Middle-aged male, FGD Upper West region

This statement clearly highlights how gendered roles mediate the destination and activity choices of climate-related migrants. As emerged from many interviews and focus groups, the preceding extract highlights changes in the agrarian sector that affect the demand for labour and consequently the patterns of migration that different kinds of migrants, including those spurred by climate change, can feed into. Beyond the mere changes in the demand for labour, the disproportionate gender effects are crucial in shaping how men and women chose different migration pathways when affected by climate change.

In the Eastern region, a traditionally nonsending region, many young people facing climate challenges in the traditional cocoa, coconut and vegetable sectors and the growing mango sector migrate mainly to urban areas to seek nonfarm jobs. The main destinations include Accra, Tema and Kumasi. 'Many young people are educated and feel that they should go to the cities to find office-type jobs. Also, they have divided all the lands among the new generations, so it is less profitable to farm unless you go deeper into the bush', explains an elderly woman from the Eastern region. Increasing educational levels and generations of land fragmentation are therefore important mediators of migration outcomes under climate challenges. Focus groups and interviews in both regions also point to a high degree of conformity to gender roles among migrants. Men and boys mostly stick to masonry, welding, driving and carpentry, while women and girls mostly participate in catering, trading, domestic work and head porterage.

Rural-rural migration

Rural-rural migration, as shown in Tables 9.3 and 9.4, is the second most important typology of migration in relation to climate change adaptation, particularly in the Upper West region. As mentioned earlier, the key driver of this process is the aim to continue pursuing agrarian livelihood strategies in the face of a changing climate. In the Upper West region, focus groups point to the growing incidence of rural-rural migration when climate-related migration occurs within the Savannah zone. The community of Siriyiri in the Wa West district is particularly noted for this trend in mobility, which mainly involves group migrations of women and children, and in fewer instances men. These groups are usually formed not on the basis of lineage, but of social networks. The groups, often led by a male, mostly migrate to meet the harvesting season of cassava and other crops in the Gonja land area (about 135 kilometres away). They spend up to four months harvesting crops with shared roles for men, women and children along traditional lines. Men are mainly responsible for the initial stage of harvesting, such as uprooting cassava, cutting down millet and maize stalks, cutting off millet crowns and plugging maize, while women and children transport these to aggregation points and are responsible for much of the further processing, such as peeling cassava and dehusking maize. The women are also responsible for preparing the group's meals.

Our FGDs emphasized that female children are preferred to accompany the groups, as they are needed to cater for their younger siblings to allow their mothers enough time for work. It was also argued that young and single females prefer southward migration to the group migrations in the local savannah, as this enables them to have autonomous control over their earnings. Group participation, on the other hand, makes it difficult to hide one's accumulated wages and gains from the relatives back home, who expect a lot of it to reach the household.

Another important new form of migration that is driven by climate change is local savannah rural-rural migration involving illegal gold-mining activities ('galamsey'). Young people migrate to communities within the Savannah such as Banda/Nkwanta, Mandari, Tinga, Nangodi and Tasilma. Respondents assert that until roughly the last two decades, 'galamsey' migrations within northern Ghana (the Savannah) were not practised in their communities. Rather, 'galamsey' migrations to southern Ghana were the norm for many decades, following the history of northern labour being recruited for the colonial mines. The 'galamsey' migrations are mainly the domain of males, with minimal participation by females:

> Galamsey has a lot of tough evil spirits, so only men can deal with them; women and girls must hide from these ones. Also, galamsey is a very tough activity, so it is for only men, and the ladies only go there either to their boyfriends or to work as secret sex workers, labourers for less intense mining processes or as cooks. So, when the rains fail, that is hardly in their migration plans. (Galamsey seasonal female migrant, Upper West)

It is also said that females are deterred from 'galamsey' migration because of the sex-work stigma attached to it. Women who engage in this migration are thus despised, being suspected of engagement in sex work. Conversely, although knowledge of men seeking sex from female workers abounds, the men who migrate are treated differently as people without a blemish. Moreover, the men are seen as local heroes who are daring to do what is unusual when they return with something to show for it, unlike the women, who are despised for it. The cultural system of male dominance which encapsulates much of the victimization of women and girls perhaps explains this situation. The literature on migrant sex workers acknowledges the existence of high levels of stigmatization and discrimination against actual and suspected women sex workers in Africa (Scorgie et al, 2013; Mgbako, 2019).

Conclusion

This chapter has highlighted the imperative of real and perceived gendered vulnerability and capability, primarily underpinned by cultural and environmental factors, in coping and adapting to climate change as a mirror to understanding the gender dimensions of climate change-related migrations. Different gender effects and the ability to adapt to climate change and gendered norms are crucial in terms of shaping how men and women chose migration pathways when affected by climate change in the Upper West and Eastern region districts.

There are relatively more people, both men and women, who are resorting to migration to deal with climate change in the Upper West region than in the Eastern region. Also, more respondents (males and females) from the Upper West region engage in seasonal migration to rural areas as well as sometimes migrating to rural areas compared to respondents in the Eastern region. In contrast, more males and females from the Eastern region sometimes migrate to urban areas, while the variations in the extent of migration in response to climate change between the regions can be explained by differential levels of vulnerability and development. As was shown earlier, the rural versus urban focus of migrations can be explained by development inequalities, changes in agrarian practices and gender norms.

The literature has generally attributed the high levels of climate-related migration from the Upper West region to the higher dependence on rainfall and higher levels of development compared to the south of the country (Ofori-Sarpong et al, 2001; Dietz et al, 2004; GSS, 2018). The climate in the Upper West region is now far less favourable, with more frequent and intense climate extremes such as droughts and floods. Moreover, lower levels of development engineered by the deliberate development policies of the colonial government and the failure of present-day governments to turn the tide have ensured that the north has remained poorer, leading to its inhabitants looking to the south for sources of livelihood such as land, especially in times of environmental shocks (Bawakyillenuo et al, 2016; Jarawura et al, 2024).

The results given in this chapter highlight how different sociocultural contexts might shape gender mobility in relation to climate change, as manifested in both new and reinvigorated forms of migration, reflecting both longstanding and transforming gendered norms. Gendered processes and conditions in societies are crucial in shaping vulnerabilities and mediating the impact of climate on migration. Furthermore, gendered roles mediate the migration pathways that climate-related migrants take up when confronted by climate crises at home. It is therefore imperative to unravel the gender dynamics of different climate-affected regions and to grasp how they mediate migration outcomes for males and females in order to arrive at an improved understanding of climate-human mobility outcomes.

References

Adepoju, A. (2010) 'Introduction: rethinking the dynamics of migration within, from and to Africa', in A. Adepoju (ed.) *International Migration within, to and from Africa in a Globalised World*. Accra: Sub-Saharan Publishers, pp 9–45.

Adger, W.N. et al (2007) *Assessment of Adaptation Practices, Options, Constraints and Capacity*. Cambridge: Cambridge University Press.

Agrawal, A., Chhatre, A. and Hardin, R. (2008) 'Changing governance of the world's forests', *Science*, 320(5882): 1460–1462.

Attua, E.M. (2003) 'Land cover change impacts on the abundance and composition of flora in the Densu basin', *West African Journal of Applied Ecology*, 4(1): 27–34.

Bawakyillenuo, S., Yaro, J.A. and Teye, J. (2016) 'Exploring the autonomous adaptation strategies to climate change and climate variability in selected villages in the rural northern savannah zone of Ghana', *Local Environment*, 21(3): 361–382.

Beine, M. and Jeusette, L. (2021) 'A meta-analysis of the literature on climate change and migration', *Journal of Demographic Economics*, 87(3): 293–344.

Bohle, H.G., Downing, T.E. and Watts, M.J. (1994) 'Climate change and social vulnerability: toward a sociology and geography of food insecurity', *Global Environmental Change*, 4(1): 37–48.

Christiansen, J.H. et al (2007) 'Regional climate projections' in S.D. Solomon et al, *Climate Change 2007: The Physical Science Basis. Contribution of Working Group I to tal Panel on Climate Change*. Cambridge: Cambridge University Press, pp 847–940.

Demetriades, J. and Esplen, E. (2010) 'The gender dimensions of poverty and climate change adaptation', *Social Dimensions of Climate Change: Equity and Vulnerability in a Warming World*, 4(2): 133–143.

Dickson, K.B. and Benneh, G. (1988) *A New Geography of Ghana*. London: Longman.

Dietz, T., Millar, D., Dittoh, S., Obeng, F. and Ofori-Sarpong, E. (2004). 'Climate and livelihood change in north east Ghana', *The Impact of Climate Change on Drylands: Environment & Policy*, 39. DOI: 10.1007/1-4020-2158-5_12

Djoudi, H. and Brockhaus, M. (2011) 'Is adaptation to climate change gender neutral? Lessons from communities dependent on livestock and forests in northern Mali', *International Forestry Review*, 13(2): 123–135.

Foresight (2011) *Final Project Report: Migration and Global Environmental Change*. London: Government Office for Science.

GSS (Ghana Statistical Service) (2012) *Population Census Report*. Accra. Ghana Statistical Service.

GSS (2018) *Ghana Living Standards Survey Round 7 (GLSS7): Main Report*. Accra: Ghana Statistical Service.

Hummel, D. (2016) 'Climate change, land degradation and migration in Mali and Senegal: Some policy implications', *Migration and Development*, 5(2), 211–233.

IPCC (Intergovernmental Panel on Climate Change) (2007) *Climate Change 2007: Impacts, Adaptation and Vulnerability. Contribution of Working Group II to the Fourth Assessment Report of the Intergovernmental Panel on Climate Change*. Available from: https://www.ipcc.ch/site/assets/uploads/2018/03/ar4-wg2-intro.pdf

IPCC (2014) *Regional Aspects (Africa). Climate Change 2014: Synthesis Report – Summary for Policy Makers.* Available from: https://www.ipcc.ch/site/assets/uploads/2018/02/AR5_SYR_FINAL_SPM.pdf

IPCC (2018) *Special Report on the Impacts of a Global Warming of 1.5°C.* Available from: https://www.ipcc.ch/sr15/

Jarawura, F.X. and Smith, L. (2015) 'Finding the right path: climate change and migration in Northern Ghana', in F. Hillmann, M. Pahl, B. Rafflenbeul and H. Sterly (eds) *Environmental Change, Adaptation and Migration: Bringing in the Region.* London: Palgrave Macmillan, pp 245–266.

Jarawura, F.X. et al (2024) '"These days, things have changed": historicizing current dynamics of climate-related migration in the savannah zone of Ghana', *Climate and Development*: 1–11. https://doi.org/10.1080/17565529.2024.2321168

Kaczan, D.J. and Orgill-Meyer, J. (2020) 'The impact of climate change on migration: a synthesis of recent empirical insights', *Climatic Change*, 158(3): 281–300.

Lama, P., Hamza, M. and Wester, M. (2021) 'Gendered dimensions of migration in relation to climate change', *Climate and Development*, 13(4): 326–336.

Lawson, D.F. et al (2019) 'Children can foster climate change concern among their parents', *Nature Climate Change*, 9(6): 458–462.

MacGregor, S. (2010). '"Gender and climate change": from impacts to discourses', *Journal of the Indian Ocean Region*, 6(2): 223–238.

McLeman, R. (2013) 'Developments in modelling of climate change-related migration', *Climatic Change*, 117(3): 599–611.

Mertz, O., Mbow, C., Reenberg, A. and Diouf, A. (2009) 'Farmers' perceptions of climate change and agricultural adaptation strategies in rural Sahel', *Environmental Management*, 43: 804–816.

Mgbako, C.A. (2019) 'Sex work/prostitution in Africa', *Oxford Research Encyclopaedia of African History.* https://doi.org/10.1093/acrefore/9780190277734.013.562

Obokata, R., Veronis, L. and McLeman, R. (2014) 'Empirical research on international environmental migration: a systematic review', *Population and Environment*, 36(1): 111–135.

Ofori-Sarpong, E. and Annor, J. (2001) 'Rainfall over Accra, 1901–90', *Weather*, 56(2): 55–62.

Owusu, K. and Waylen, P. (2009) 'Trends in spatio-temporal variability in annual rainfall in Ghana (1951–2000)', *Weather*, 64(5): 115–120.

Popoola, O.O., Monde, N. and Yusuf, S.F.G. (2018) 'Perceptions of climate change impacts and adaptation measures used by crop smallholder farmers in Amathole district municipality, Eastern Cape province, South Africa', *GeoJournal*, 83(6): 1205–1221.

Riede, J.O., Posada, R., Fink, A.H. and Kaspar, F. (2016) What's on the 5th IPCC Report for West Africa? *Adaptation to climate change and variability in rural West Africa*, pp 7–23.

Scoones, I. (1998) *Sustainable Rural Livelihoods: A Framework for Analysis*. IDS Working Papers 72. Brighton: Institute of Development Studies.

Scorgie, F. et al (2013) 'Human rights abuses and collective resilience among sex workers in four African countries: a qualitative study', *Globalization and Health*, 9(1): 33.

Shackleton, S., Ziervogel, G., Sallu, S., Gill, T. and Tschakert, P. (2015) 'Why is socially-just climate change adaptation in Sub-Saharan Africa so challenging? A review of barriers identified from empirical cases', *Wiley Interdisciplinary Reviews: Climate Change*, 6(3): 321–344.

Teye, J.K. and Owusu, K. (2015) 'Dealing with climate change in the coastal Savannah Zone of Ghana: in situ adaptation strategies and migration', in F. Hillmann, M. Pahl, B. Rafflenbeul and H. Sterly (eds) *Environmental Change, Adaptation and Migration: Bringing in the Region*. London: Palgrave Macmillan, pp 223–244.

Teye, J.K., Jarawura, F., Lindegaard, L.S., Kleist, N., Ladekjær Gravesen, M., Mantey, P. and Quaye, D. (2021) *Climate Mobility: Scoping Study of Two Localities in Ghana*, DIIS Working Paper, No. 2021: 10. Copenhagen: Danish Institute for International Studies.

Tufuor, T., Sato, C. and Niehof, A. (2016) 'Gender, households and reintegration: everyday lives of returned migrant women in rural northern Ghana', *Gender, Place & Culture*, 23(10): 1480–1495.

Van der Geest, K. (2011) 'North-south migration in Ghana: what role for the environment?', *International Migration*, 49(1): 69–94.

Warner, K. and Afifi, T. (2014) 'Where the rain falls: evidence from 8 countries on how vulnerable households use migration to manage the risk of rainfall variability and food insecurity', *Climate and Development*, 6(1): 1–17.

Wrigley-Asante, C., Owusu, K., Egyir, I.S. and Owiyo, T.M. (2019) 'Gender dimensions of climate change adaptation practices: the experiences of smallholder crop farmers in the transition zone of Ghana', *African Geographical Review*, 38(2): 126–139.

Yaro, J.A. (2013) 'The perception of and adaptation to climate variability/change in Ghana by small-scale and commercial farmers', *Regional Environmental Change*, 13(6): 1259–1272.

Yaro, J.A., Teye, J. and Bawakyillenuo, S. (2015) 'Local institutions and adaptive capacity to climate change/variability in the northern savannah of Ghana', *Climate and Development*, 7(3): 235–245.

Zhao, X., Akbaritabar, A., Kashyap, R. and Zagheni, E. (2023) 'A gender perspective on the global migration of scholars', *Proceedings of the National Academy of Sciences*, 120(10): e2214664120.

10

Trespassing Legal and Moral Boundaries: Ethiopian Domestic Workers Returned from the Middle East

Ninna Nyberg Sørensen

Introduction

> What can we do against these human traffickers if women want to go?
> Labour and Skills Officer, Shashemene town

In the Labour and Skills Office in Shashemene, the head of office has agreed to meet with the project team to give an update on migration as it exists in the region. The office is tasked with local job creation, but increasingly also with preventing 'illegal migration' and human trafficking. For example, it launches information campaigns about the risks involved in irregular migration and provide prospective migrants with the skills that will presumably enable their 'migration through legal means', the latter being understood as bilateral agreements that respect migrants' rights. To our question about challenges in that regard, the officer shrugs: 'We have discussions with returnees, most of whom are women who have come back because they were deported or because they were suffering abuse from their employers. No matter what the reason, most plan to go again.' Government measures against irregular recruitment agencies, primarily by cancelling licenses, have had only a limited effect as new agencies simply take over. In Shashemene more than 22 agencies are facilitating migration to the Gulf and the Middle East, among which many, according to the official, use 'illegal means that put people's lives at risk'. But what about assistance directed at returned migrants? We are offered an example of a project that has provided shades with full equipment for setting up roadside shops for soda vending. Each shade is intended to provide a selling spot for five returnees. Apart from the Coca Cola company, which donated the shades, three nongovernmental organizations (NGOs) are involved, one as a financial institution providing loans to set up shop. The NGOs request a business plan, and the office approves and signs off the plans. So far, some 40 soda-vending stations have been approved.

Across town, the Women and Children's Affairs regional office assists prospective migrants who wish to work abroad with passports and training. The office is also tasked with the rehabilitation of deported youth. 'We struggle to stop this traffic in women', the newly appointed head of office explains: 'We want women to have the knowledge and skills that can enable them to live their lives here.' However, there are few means to do so. The government provides the building and the officer's salary, but does not allocate resources for activities. The officer's main job consists in writing project-funding applications to NGOs. Over the last couple of months, 250 returned women have registered with the office. Only ten have received assistance. Despite constant evictions from the Gulf States, the office is currently without any project money. People register, some with physical injuries sustained abroad. Most walk away frustrated.

Representatives of parallel institutions in Haiq and Kombolcha are also frustrated. Given that North and South Wollo are the main sending areas for migrants destined for the Gulf and the Middle East, it is a matter of concern that no specific government programme targeting returned migrants is currently running. Referring specifically to migrants who returned during and due to the COVID-19 pandemic, a representative speculates that 'return migrants may be more desperate than other unemployed youth'. Many, especially the regular domestic workers, were returned before their labour contracts expired. Some lost their income and savings during the time spent in detention camps or in quarantine in Addis Ababa and were only reluctantly received back by their families, who had lost investments on what they had imagined to be a better future. Officials agree that, regardless of the hardships experienced, most returnees will probably leave as soon as an opportunity presents itself.

These narrations of the current migration situation are striking in several ways. One cannot ignore the deeply felt desire to provide much-needed assistance to prospective, returned or remigrating nationals, or the frustration over a less than adequate allocation of funds with which to offer rehabilitation, reintegration or simply alternatives. However, what is equally hard to miss is the way in which the language of stricter migration management – including preventing irregular migration in different forms – spills over into the way in which migration is articulated. It is a difficult task to stop human smugglers and traffickers when women are only awaiting their next chance to leave the country. However, as women are presumed to be more vulnerable to human trafficking than men, protecting them from international migration is apparently for their own good.

Danger lurks outside the walls of the home and the national space. Trafficking, abuse and exploitation are common concerns in discussions of Ethiopian migrant domestic workers, and the risks involved in travelling irregularly are well documented.[1] So are the poor living and working conditions that domestic workers encounter abroad, including long working hours, physical and sexual abuse and even assault, the withholding of passports and the refusal to pay salaries.[2] In the first case, attention has centred on deceitful recruiters, unscrupulous human smugglers and violent human traffickers, who are believed to form part of international criminal networks. The combined effects of economic hardship, protracted conflict and climate-change hazards are predicted to increase the risk. While these are legitimate concerns, deeper considerations of why and how Ethiopian women seek out migration in the first place are often lacking. Is it possible that women consider their chances for improvements in their relative wellbeing to be higher through migration than through locally available avenues? Could it be that they are aware that successful migration takes hard work, perseverance and the endurance of a degree of hardship, which they hope will be of a temporary nature? In the absence of decent local jobs and adequate provisions for social protection, migration often becomes the only viable means of alleviating risks. Focusing only on the risks faced during irregular journeys and potentially exploitative working conditions easily overlooks the risks involved in staying put in their country of origin. In comparison to the risks of environmental degradation, equally exploitative local jobs or various levels of conflict, irregular migration may offer the possibility at least for higher earnings that can shape the future in more promising ways.

This chapter explores the structural and social circumstances that prompt Ethiopian women to engage in a variety of mobility practices. It also discusses the political and sociocultural factors that circumscribe how women's mobility is conceptualized and talked about in Ethiopian society and international policy circles. As will become apparent, women's irregular migration is often referred to as human trafficking without it necessarily being so.

To avoid predetermining a multitude of experiences as necessarily human trafficking and opening our eyes to women's often restricted agency, the chapter introduces the concept of trespassing. Albert Hirschman (1981) conceived of the idea of conceptual trespassing in the field of development studies as an analytical means of questioning conventional stances and understanding new dynamics. Pucci and Vecchio (2019) have recently suggested trespassing as a relevant conceptual, analytical and operational approach to mobility studies. Most people associate trespassing with an unauthorized entry on someone else's property. That meaning is also part of the *Meriam-Webster Dictionary* definition of the noun 'trespass'. The other part is 'a violation of moral and social ethics'. Used as a verb, as in 'to commit a trespass', the *Meriam-Webster Dictionary* likewise suggests a dual definition: either 'to enter unlawfully upon the land of another' or 'to violate the bounds of good behaviour or taste'. Both meanings prove useful for the following analysis.

Ethiopian migrant women cross various boundaries, including state borders and gendered norms for proper behaviour, when travelling and working abroad. Their trespassing makes them vulnerable to exploitation and human trafficking, but may simultaneously be understood as a way of tricking migration control or overcoming other obstacles surrounding their mobility. The chapter suggests using trespassing as an analytical devise to capture the possibility of multiple outcomes.

The chapter is organized as follows. The next section provides an overview of migration routes, particularly those of Ethiopian migrant domestic workers, the system governing their mobility and the vulnerability to human trafficking that haunts the system. The following section provides a brief methodological background to the empirical findings, as well as some reflections on the context in which a great deal of contemporary human trafficking research is conducted. This is followed by a section presenting and discussing the findings, on which the final section then concludes.

Migration for domestic work in the Middle East

The experiences of Ethiopian migrant domestic workers in the Middle East have received significant attention, including what happens along the routes people travel to reach their destinations, the working conditions they face upon their arrival and the lives they live on the legal margins of society if they are relegated to (or opt for) irregular migration. Considering that a majority are believed to travel through irregular channels, reliable international migration statistics for Ethiopian women are hard to find (Gezahegne and Bakewell, 2022). It is commonly believed that Ethiopia's international migration rate remains lower than the Sub-Saharan average of 2.5 per cent, that it nevertheless has increased since the beginning of the 1990s and that it is likely to grow further in the years to come (Asnake and Fana, 2021). Ethiopian migrants tend to travel along three routes: east to the Persian Gulf states and the Middle East, south to South Africa and north(west) into Sudan and, if the aim is reaching Europe, onwards to Libya (Adugna, 2021). Women travel along all three routes, but mainly head to the Gulf States and Middle East, where for years they have found employment as domestic workers.[3]

The 'kafala' system

Jordan, Lebanon and all the Arab Gulf states except Iraq and, since 2018, Qatar[4] regulate migrant residence and employment through a sponsorship system known as 'kafala'. A local citizen or company (the 'kafeel') must sponsor any foreign worker for their work visas and residence to be valid, cover the travel and living expenses of their workers, and pay the agreed wage. Insufficient regulations and legal protections for migrant workers' rights leave the workers at the mercy of the sponsor, potentially resulting in low wages, poor working conditions and vulnerability to abuse (Robinson, 2021). Racial discrimination and gender-based violence are reported.[5]

Feminist scholarship has pointed to three critical functions of the kafala system that deserve mentioning: first, the state effectively decentralizes and privatizes regulatory functions to the sponsor (Fernandez, 2011); second, the system offers private actors a lucrative business opportunity to capitalize on the demand for work visas (Longva, 1997); and, third, the cultural norms that protect the privacy of the home from intrusion by the state have not only made the inspection of domestic workers' labour conditions difficult but have also effectively masked the lack of public provisions for social reproduction in countries of destination, as functions such as childcare and care of the elderly are provided by a docile and low-cost workforce (Fernandez, 2021), while a lack of social protection is compensated for by women's remittances in countries of departure.

Ethiopia has recently come under pressure to protect national migrant workers and international pressure to halt irregular migration. Since the partial ban on women travelling to Lebanon in 2008 and the 2013 five-year ban on domestic workers moving overseas, Ethiopia has signed bilateral labour agreements with Jordan, Qatar, Saudi Arabia and the United Arab Emirates (Adugna, 2021). Apart from the unintended consequence of spurring more irregular migration, it could be argued that in doing so, the Ethiopian state has drawn a curtain over a lack of public provision for social protection, as this problem is solved by its growing out-migrating population. The more than 100,000 Ethiopians detained for subsequent deportation after an agreement made between Saudi Arabia and Ethiopia on 30 March 2022[6] indicate state complicity in upholding the system. So does the March 2023 announcement of 'the opportunity of a lifetime' for some 500,000 Ethiopian women to become domestic workers in the Gulf, for which the Ethiopian government has been criticized for seeking to ease national unemployment rates and make migrants contribute financially to the country's economy through the sending of remittances.[7] Thus, this chapter argues that any analysis of migration governance must inevitably include not only the migrant-receiving state, but also the sending state's authorization of such bilateral agreements.

Human trafficking

Trafficking in human beings refers to the recruitment, transportation, transfer, harbouring or receipt of persons by means of the threat or use of force or other forms of coercion. To qualify as human trafficking, three elements must be present: (i) the act of transferring, transporting, or recruiting a person; (ii) using the means of threat, coercion, force, deception or abduction to control the person; and (iii) with the purpose of forced labour, sexual exploitation or slavery (Beck et al, 2017).

The US Department of State's Office to Monitor and Combat Trafficking in Persons has for many years identified Ethiopia as a country with a burgeoning human-trafficking problem, be it for adult or child labour, forced prostitution or organ harvesting. The Government of Ethiopia does not fully meet the minimum standards for eliminating human trafficking, but it has made progress, resulting in an upgrade from the Tier 2 Watch List to Tier 2 in the 2023 TIP Report (US Government, 2022, 2023). Women are reportedly at a higher risk due to their lower status in the family, community and society (Gezie et al, 2021).

As accurate data on the extent of trafficking in human beings are hard to find, human trafficking is believed to remain 'under-reported, under-detected and under-prosecuted' (Goodey, 2008: 425). In the Ethiopian case, overreporting may constitute a different, albeit related problem. Empirical studies based on a relatively few cases may have contributed to generalizing the idea that most migrant women who travel irregularly fall prey to traffickers and experience abuse (for example, Reda, 2018), while scoping studies identifying risk factors may be taken as evidence for the prevalence of the phenomenon (Beck et al, 2017; Demissie, 2018; Gezie et al, 2021). Additional risk factors attributed to the Ethiopian state include corruption, instances of violence and abuse committed by border police, weakly enforced laws, inconsistent prosecution, inconsistent application of public policy, inadequate protection services and a lack of coordination between different authorities.

However, the increasing illegalization of a range of mobility practices risks leading to both a reflex-categorization of women's irregular migration as human smuggling or trafficking (Ayalew, 2019; Cockbain and Bowers, 2019; Adugna et al, 2021), as does an analytical perspective grounded in the 'migration-crime-security-nexus' (Goodey, 2008).

An alternative analytical framework: structural, political and domestic violence

The forces driving Ethiopian women's irregular migration may perhaps be better understood as embedded in the broader economic, political and demographic transitions that occur across Ethiopia. Apart from rapid economic growth and a restructuring of the country's internal administration along ethnolinguistic lines (resulting in internal displacement), a high fertility rate translates into a youthful population whose transition from childhood to adulthood becomes delayed due to a high youth unemployment rate (Kefale and Gebresenbet, 2021). The delay in the transition from youth to adulthood is a result of several factors, including prolonged time spent in the educational system, shortages of arable land and insufficient employment opportunities. As a result, migration has become one of only a few available options for transiting from childhood to adulthood.

Research conducted by Kerilyn Schewel (2022) demonstrates that Ethiopian women's migration to the Gulf States can be a 'reasonable capability-enhancing choice' in response to a critical lack of local pathways to progress or personal life aspirations. Migration often arises at a particular moment in a woman's life course, for instance, as adolescents transition into adulthood. In such instances, migration represents a way 'to disrupt an otherwise predictable future' (Schewel, 2022: 1628). Migration trajectories are influenced by a multitude of contexts and objectives. Young women migrate both as part of wider family decision making and a desire to live new lives on their own terms away from social constraints (Grabska and de Regt, 2022). Adult women's migration may be a response to social events such as divorce, widowhood or single motherhood. The transition perspective provides a way to foreground the fact that the reasons driving women's irregular migration may be more complex than accounts of human trafficking suggest.

Could women's migration be grounded in a wish to escape structural, political and domestic forms of violence, as well as in desires for opportunity and autonomy? According to Lauren Carruth and Laura Smith (2022), such desires go well beyond financial gain and may include having a better life, becoming independent from one's family, being able to establish a home of one's own or overcoming different forms of gender-based violence. Their interviews with irregular Ethiopian women migrants from the Oromia region revealed that a combination of protracted violence in the women's communities, threats and harassment faced by themselves or close relatives (including the confiscation of land and the arrest and detention of local activists), as well as intimate partner violence figured as prominently, as did economic insecurity in motivating women's journeys.[8]

In rural areas, violence against women and children is widespread (Semahegn and Mengiste, 2015). Being divorced or poor are additional predictors (Chernet and Cherie, 2020). Murphy et al (2021) asserts that some parents find it acceptable to punish adolescents physically if they do not comply with unpaid labour demands. Girls are mainly punished for violating gender norms. During periods of prolonged drought, the drivers of violence increase; the longer distances women and girls must walk to fetch water may become sites of attacks. The nature of violence varies considerably across regions and religious groups, for example, regarding the prevalence of early or arranged marriage, or polygamy.[9] So do tactics to escape it. Meron Zeleke

Eresso (2019) has pointed to the practice of migrant sponsorship among sisters as a way of fleeing different forms of gender-based violence. As we will see later on, migration organized through women-centred networks may prove the safest way of trespassing gendered moral boundaries.

Ethiopian women face disproportionate levels of violence when navigating multiple concurrent transition processes and legal spaces. Nevertheless, the dominant discourse of trafficking leaves little room for discussing the extent to which domestic, political or structural violence within Ethiopian society may precede migration decisions (Carruth and Smith, 2022). Taking account of both structural and domestic forms of violence, as well as the tactics to escape it, I argue in favour of locating Ethiopian women's migration experiences in the overall context of multiple overlapping realities.

Marina de Regt (2010) has suggested that gender, agency and (il)legality play a role in all phases of the migration trajectory: when leaving Ethiopia, when travelling and entering the destination country, when residing and working abroad, and when eventually returning (or moving elsewhere). The relationship between way of travel, mode of entry and legal status is never fixed. Rather, women 'move in and out of il/legality', sometimes by choice, as illegality may present them with fewer duties, such as leaving control to sponsors or having to comply with conditions set out in legal contracts (de Regt, 2010: 240). Although social relations between migrants, brokers and employers remain asymmetric, they come in many forms and may be resisted in different ways. I draw on the insight that illegality not by default is disadvantageous in the analysis.

Methodological reflections

This chapter relies on data collected during 2021 and 2022 in two distinct areas of Ethiopia: the Shashemene woreda in the West Arsi zone of the Oromia Regional State, and the Tehuledere woreda in the South Wollo zone of the Amhara Regional State.[10] The Shashemene interview material consists of a total of 38 interviews carried out among representatives of state offices, the kebele administration and NGOs, community elders, community members of different genders and ages, domestic migrants and return migrants. In addition, 11 focus group discussions (FGDs) were organized with farmers (mainly men), women and youth groups from Shashemene woreda and the rural kebeles of Faji Sole, Haleche Harabate and O'ene Chefa Umburu. In Tehuledere, ten return migrants (nine women and one man) and heads of households with return migrants were interviewed. In addition, two local government representatives were interviewed in Kombolcha town and Haik.

The COVID-19 pandemic affected work undertaken in both field sites, as did political instability and conflict, the latter primarily limiting access to the Tehuledere field site for follow-up interviews. Household surveys and FGDs were initially conducted in both sites, coordinated by senior researchers Zerihun Mohammed (with the assistance of Busha Teshome and Beriso Tufule, Shashemene woreda) and Dessalegn Rahmato (with the assistance of Andnet Gizachew, Tehuledere woreda). Additional follow up-interviews at the woreda and kebele levels in Shashemene were carried out during February 2022 by the author, Neil Webster, Zerihun Mohammed and Adane Alemayehu, with translation provided by the last two. We talked to regional state representatives, community elders, kebele chiefs, groups of farmers, women workers in the industrial park in Hawassa and returned migrants. Interviews and more informal conversations with return migrants and migrant households concerned themes such as how climate change affects local livelihoods, where family members find employment, how migration is organized and paid for, family and community expectations of migration and return, and migrant working conditions and earnings. Interviews with government agents touched on observable environmental change, the general migration situation in the area, attempts at regulation (including of brokers), alternatives to migration and the assistance provided to returned migrants. Interviews generally lasted around an hour.

The women migrants we talked to in Shashemene had mostly returned prior to the COVID-19 pandemic, whereas migrants interviewed in Tehuledere were returned due to the pandemic. In both settings, qualitative interviewing was exploratory with an eye to producing exemplary data and capturing diversity in experiences and research participants' interpretations of these. Following field visits, additional insights were gained through studying the websites and videos uploaded to YouTube by Ethiopian migrant domestic worker-led activist groups and NGOs offering shelter and rehabilitation to human trafficking survivors.[11]

Lauren Carruth and Laura Smith (2022) note that much scholarly and NGO-based research with migrants is retrospective and based on conversations with return migrants, whose relative 'successes' and 'failures' may shape their memories and responses. When it comes to experiences of human trafficking, yet another problem is that our knowledge 'is largely based on qualitative victims' accounts gathered by NGOs, which actively assist victims' (Goodey, 2008: 426). Although these organizations are often well placed to estimate the extent of the problem, we should not neglect that their claims of a growing human trafficking problem have resulted in steadily increasing resources being allocated to anti-trafficking efforts and thus also to the growth of the anti-trafficking industry. Additional critiques of notions of human trafficking being tied to political outcomes have been raised by Harmon et al (2022) and Horwood et al (2022).

We did not approach return migrants through organizations assisting survivors of human trafficking for our research. This is an important disclaimer, as it could partly explain why clear-cut human trafficking experiences are not prevalent in our findings. Several interviewed women had experiences with human smugglers, but not necessarily in exploitative roles.

A final disclaimer regards the site-bound bias of basing the analysis on interviews within Ethiopia mainly with return migrants and officials. Successful migrants might not have gone back, at least not to their communities of origin. Further, the focus in recent years on combating irregular migration inevitably sets the tone for how things are talked about in public. As pointed out by Adugna et al (2021), a high number of anti-human trafficking taskforces have been established at all tiers of the government, from the office of the Deputy Prime Minister to district levels all over the country. The structure allegedly extends to the village level, where the community, elders, religious leaders and clan heads have been organized to prevent 'illegal migration' and instructed to report any act of migration facilitation, including the families and brokers who organize departures and smuggling in the border towns. At the woreda level, we found a discrepancy between officials' claim that up to 25 per cent of the households had a migrant abroad, whereas relatively few parents wished to talk about having migrants within their own family. Apart from the fear of oversight by local taskforce members, access to or qualifying for humanitarian or social assistance may be a reason for withholding information on having a son or daughter abroad.

When women go abroad

As the previous discussion has indicated, Ethiopian women's migration is conditioned by a range of interlinked factors, including endemic poverty, environmental degradation, recurrent famine, poor governance, limited social protection provision, political unrest and vulnerability to gender-based violence (including in the family). Rather than predetermining their irregular mobility as human trafficking, a scaled understanding based on the idea that the point at which tolerable conditions end and human trafficking begins might bring new insights.

At one end of the scale, we would find women who have been thoroughly deceived about travel and working conditions, are confined to the domestic workplace, especially when travelling and working within the kafala system, fall into debt bondage and are physically or psychologically abused. At the other end are women who operate with some knowledge and agency, and who are not necessarily deceived by brokers or mistreated by employers (Weitzer, 2014; Sørensen, 2019). Ethiopian women migrants often change position during their migration trajectory. To catch such dynamics, conventional synchronic approaches must be supplemented with studies that take a more longitudinal look.

At the local level, people speak of human trafficking as something to be wary of. Three young women workers in Hawassa industrial park told that one of the reasons they had chosen not to go to the Middle East was the stories they had heard about hazardous journeys and 'what happens to girls there'. Village elders and other local authorities also spoke about the risks involved, but underscored the 'unlawfulness' of travelling without permission. The return migrants we talked to had mainly used irregular channels, some in combination with regular travel arrangements, either by entering Saudi Arabia on a tourist visa obtained for the purpose of pilgrimage to Mecca that later was overstayed, or by breaking an authorized kafala contract to begin working independently. The general picture is thus one of moving in and out of (ir)regularity. Several of our informants pointed to irregular travel arrangements as both cheaper and faster than government-sanctioned kafala contracts. Travel costs were reported that are as low as 10,000 ETB (around 173 EUR) in Tehuledere, but up to 70,000 ETB (1,200 EUR) in Shashemene, where formal agencies abound. In Tehuledere we were told about two cases of deceit, confinement and demands for extra payments along the route from non-Ethiopian smugglers. However, we also found that people used the terms 'brokers', 'smugglers' and 'traffickers' somewhat interchangeably. Most talked about 'delaloch' (brokers), but also used 'mengdemerwecho' (literally road guide) and other local terms.

Both locals and returnees referred to labour exploitation in terms of bad treatment by employers, primarily related to long working hours. In both regions, people knew of women returning with long-term depression or severe psychological problems, but also of women who had decided to stay on and endure sometimes harsh conditions because of the salary and, at least during the initial hard time abroad, their need to pay back the travel debt.

We were told that competition for land may make the boys stay around, whereas the girls can be sent abroad. Lower yields from agriculture caused by declining or unpredictable rainfall are making rural communities and families more dependent on additional cash incomes. Some believed that many families push their young members to migrate, even though it may be more difficult to marry off a daughter after her eventual return. Women related the ever-longer distances they must walk to find water due to drought and pointed to these distances as a reason behind seeking employment abroad. Both community elders and government representatives talked about women's migration as leading to 'family problems', family disintegration or conflicts between husbands and wives, often followed by court cases. While it may be difficult to establish direct causal relationships, the experiences presented in what follows suggest that some of the family problems referred to may precede the migration rather than be a consequence of it.

Moving in and out of legality

Rahima[12] returned to Shashemene from Kuwait in 2020, incidentally three weeks before anyone had heard of COVID-19. She came back 'exhausted'. Her migration experience included two years in Dubai – 'no good, only small money' – a return to Ethiopia, remigration to Kuwait for three years, another return to Ethiopia, and finally two more years in Kuwait, until her recent return, which she considers final. Prior to migration, she sustained a living as a market vendor, but had trouble feeding her two children after her first husband died and a second husband did not contribute financially. She decided to go abroad to support her two children, at that time aged 11 and 12. Travel arrangements for Dubai were made through official channels; the two stays in Kuwait were irregular. In fact, Rahima considered irregularity a better option: 'Illegal agencies in Kuwait offer jobs to women without papers. It is better to be without. The salary is higher.' Thus, trespassing legal boundaries meant better earnings. Despite the hardships, Rahima underscores that migration enabled her to give her children a better education. But she also acknowledges her lack of control over the substantial sums remitted to her family, which had been intended for buying a plot of land and building a house on it. That she found an empty plot upon her return can hardly be blamed on human trafficking.

Deceived by employers, relatives or both?

Momina embarked on an Umrah pilgrimage to Saudi Arabia in 2003. Relatives in Jeddah organized her travel and first domestic employment. Communication turned out to be a major problem. Work was heavy and she didn't speak Arabic. Apart from taking care of the family's six children, Momina also had to clean. One day, when she went out with the rubbish, the children locked the door: 'I couldn't tell them to open, because I didn't speak their language', she recalled. After staying outside in the extreme heat for six hours, her employer came home and scolded her for not having taken care of her chores. She cried and could not face the often 20 hours of daily work, as she also had to work for her Ethiopian relatives to repay the favour of facilitating her migration. 'This is how you have to work in Saudi', her relatives said. She decided to escape, left all her belongings and the salary she was owed behind, and found employment on her own, for a couple who 'paid my salary properly'. Two years prior to her return, she married a fellow Ethiopian migrant with a residence visa. They both ended up being deported during the mass deportations in 2013–2014, Momina after two months of confinement in a detention centre: 'I came back pregnant with no savings.' The money remitted to her natal family over the years had evaporated in 'unwise spending'.

Migration as a capability-enhancing choice

While family members used remittances received for other purposes than intended in the examples given earlier, other migrant women made savings and invested in their own futures. Fetiya completed 10th grade with the intention of pursuing higher education. However, a 'family problem' occurred and she wanted 'to help out'. In 2008 a regular opportunity to go to Dubai presented itself. She left her rural community and stayed on for eight years. Money earned during the first half of her stay was used to pay off debts and sustain her natal family. After that, she made her own savings. Eight years later, she thought it would be better to go back 'while still in good health'. She returned in 2017, married shortly after her return, and today has a five-year-old son. She contributes to the family income by running a small kiosk. Owning a house in Shashemene town with a business attached would not have been possible without migration.

Return during COVID-19

Tenaye directed her frustration at the Ethiopian authorities. She left Tehuledere for Dubai in 2012 'the legal way'. The travel cost (10,000 ETB, approximately 75 EUR) was paid by her family. Over the next nine years, she worked for four different employers, all 'good families'. Initially, the work was hard, but it became easier as her Arabic proficiency improved. The salary also got better. Over the years, most of her earnings were sent back to support her family, who thanks to her have become better off. When the COVID-19 pandemic reached Dubai, she decided to return to Ethiopia. Fuelling her decision was the fact that employers were restricting foreign domestic workers' mobility. They were no longer allowed to leave the house for smaller errands: 'We couldn't even go to the "Ejaza" houses where Ethiopian domestic workers spend their monthly free time and communicate with each other.' On top of feeling confined, the employers' home quarantine exacerbated the workload. She had been told that the Ethiopian authorities would cover the quarantine costs upon arrival, which turned out not to be the case. On the contrary, after a full day spent waiting in Bole airport, she was driven to a quarantine centre, where she had to spend first 14 days in the overcrowded premises, then another seven days due to a fellow returnee's positive test result. She was then told to bring a family member resident in Addis Ababa as a guarantee before leaving the Centre. On top of the quarantine costs (13,000 ETB, approximately 97 euro), the price of the air fare was at least 50 per cent above the normal price due to the pandemic. At the time of the interview, Tenaye had been back for a full year, unable to find a local job. As part of an International Organization for Migration (IOM) programme directed at youth unemployment, she had attended a training course in apparel fashion designing, but otherwise received no support. Although she would rather stay in Ethiopia, she feared that her dire economic situation would force her to remigrate.

Shala, a young woman from Tehuledere, worked for almost four years as a housemaid in Saudi Arabia before being deported due to her irregular status. Her family had covered the initial travel costs, which were set at 80,000 ETB (approximately 600 EUR) by the local broker: 40,000 ETB (approximately 300 EUR) up front, the rest upon arrival. Deceitful smugglers in Hayu (Djibouti) nevertheless requested another 80,000 ETB, threatening to imprison her unless the money was paid. Her family paid up, and she travelled on through Djibouti and Yemen to Saudi Arabia with other migrants. While there, she worked in two houses. With a monthly payment of about 1,000 Riyal (approximately 248 EUR), she was able to pay off the family debt and make some savings: 'I decided to go home because of the Saudi government's decree to deport migrants due to COVID-19. Once I recognized that they would force us to go home, I applied for voluntary and safe return.' She was immediately taken to a detention camp, where she was kept for over two months: 'It was miserable, there was not enough food and water, the facility was despairing, but thanks to Allah we survived and arrived home.' The deportation costs, including the air fare, were covered by the Saudis. Coming back early during the pandemic, Shala was not quarantined upon arrival. After having her temperature taken at the airport, she travelled on to Tehuledere. Here several people feared that she would 'bring the virus'. At the time of the interview, Shala was staying in her parent's house with her son. Feeling herself to be a burden, she wished to remigrate.

Family members in general and fathers in particular sometimes saw their daughters' return as a failed investment. The father of a young woman migrant in Kombolcha told us that 'I was not happy when my daughter returned, although her mother welcomed her. Had it not been for COVID-19, she could have supported us. If I got the opportunity, *I would send her back*' (emphasis added). Another father in Kombolcha elaborated: 'Of course, we are disappointed by our daughter's return. Not least because of the loan we squandered on her travel arrangements. She didn't even manage to earn enough to cover that.' Then he speculated: 'Had she travelled through the legal route, we might not have lost the money. They could not deny a legal migrant her salary, could they?'

Despite COVID-19, some women managed to stay abroad. A single mother told that her daughter had been in Saudi Arabia for almost two years. The idea of bringing her there was her aunt's, who said that 'she would be able to support us by going there'. She covered the (irregular) travel expenses and found the daughter a job: 'Thanks to Allah, I didn't suffer any costs. As she has been able to stay in Saudi Arabia, I am not saddened.' In this case, the trespassing of legal boundaries by women-centred family networks has not only presented a capability-enhancing opportunity but also a way to avoid deportation.

Conclusion

Gender affects the way in which migration takes place, the risks faced prior to, during and after migration, and the employment and protection options available. In Ethiopia, the combined effects of long-term poverty, recurrent and protracted political conflict, and the impact of environmental change increase women's vulnerability to human trafficking, but do not automatically and necessarily lead to it. During migration and upon their return, women who are forced to mobilize their existence by crossing borders may not immediately identify conditions associated with climate change or conflict as what had encouraged their migration, but may instead refer to deteriorating economic conditions, loss of crops, jobs and dwellings, and the debts incurred to cover such losses. Vulnerability seems to increase when mobility decisions are taken on their behalf.

Yet, as we have also seen, migration policies designed by sending and receiving states not only have direct consequences for how migration unfolds, but also have unintended consequences. In many instances, vulnerability to human trafficking may rest as much on the fact that legal options for migration are limited (as those available are tied to bilateral kafala agreements) as on abuse suffered at the hands of human smugglers and traffickers. When abuse related to travel arrangements occurs, both officials and relatives may be involved, but when abuse related to working conditions is present, women are sometimes capable of trespassing legal boundaries and finding better working conditions. Studying their experiences over time allows us to see the creative strategies they employ in negotiating precarious positions.

Why is this important? The trafficking label has severe consequences for migrants and any attempt to assist them throughout their trajectories. Its politicization has led to a systematic neglect of the significant exploitation of migrant labour. There is therefore a need to reorient the analysis of Ethiopian migrant domestic workers beyond questions of legality. This is particularly important when women are navigating poverty and precarity within informal economies. Women migrating for domestic work are often excluded from stable employment alternatives at home. Deploying an analytical lens that allows for agency and applying an economic critique of poverty as historical, structural and gendered enables an understanding of domestic work abroad as a possible way of trespassing structural constraints.

By moving the analytical focus on legality from labour to the international governance of migration, the chapter also reveals how international attempts to 'tackle migration upstream' has resulted in an effective 'seepage downstream' of the human trafficking label in which public narratives intersect with the globe-spanning governance and control of migration. The progressive externalization of the European Union's southern migration frontier is underpinned by various containment logics, including a redefinition of successful development as 'combating the root causes of migration', but also a categorization of potential and actual migrants as people at risk with the ultimate purpose of crafting sedentary subjects (Landau, 2018).

Perhaps most importantly, the chapter has elucidated the diversity in Ethiopian women's migration experiences. Some have been thoroughly deceived, confined to exploitative workplaces and subjected to various forms of abuse including from their own families, while others have taken the best route available to better their situation. Interestingly, the experiences almost always involve trespassing over legal and moral boundaries and are often intimately connected to ineffective state policies, poor service provision, unfavourable environmental conditions, political instability and gendered violence.

Notes

[1] See, for example, IOM (2021) or Zaazaa (2022), but also scholarly assessments by Ayalew (2019), Asnake and Fana (2021), Gezie et al (2021), Kefale and Gebrensenbet (2021) and Shewamene et al (2022).
[2] See material produced by activist groups such as Egna Legna Besidet or MESEWAT, media coverage like Makooi (2020), a report by the ILO (2021), and further discussions in Fernandez (2011), Fernandes and de Regt (2014), Beck et al (2017), Demissie (2018), Henry (2018), Reda (2018), Ayalew et al (2019), Nisrane et al (2020) and Zaazaa (2022).
[3] Focusing on the increased feminization of Ethiopian women's migration to South Africa, Etifanos and Freeman (2022) argue that, in this migration stream, few women migrate independently, as most are sponsored by a husband. This leaves women less room for economic independence, emancipation and empowerment.
[4] Qatar officially announced the abolition of the sponsorship system on 12 December 2016 and implemented new migration labour laws in 2018 (Al Hammadi et al, 2022).
[5] See https://humantraffickingsearch.org/the-kafala-system-an-issue-of-modern-slavery/
[6] At least 38,000 were deported over the summer of 2022. See https://www.iom.int/news/funding-needed-assist-over-100000-ethiopian-migrants-returning-kingdom-saudi-arabia
[7] See https://www.aljazeera.com/features/2023/4/17/ethiopia-recruits-500000-women-for-domestic-work-in-saudi-arabia
[8] The nature of gender-based violence, such as partner violence and early or arranged marriage, varies across regions in Ethiopia.
[9] Personal communication with Meron Zeleke Eresso.
[10] Ethiopia is administratively divided into four levels: regions (regional states), zones (subdivision of regions), woredas (districts) and kebeles (wards or the lowest tier of the government administrative system).
[11] See www.mesewat.org, www.egnalegna.org, https://agarethiopia.org and https://oprifs.org.et
[12] Rahima and all other names used in the text are pseudonyms.

References

Adugna, G. (2021) 'Once primarily an origin for refugees, Ethiopia experiences evolving migration patterns', *Migration Information Source*, 5 October 2021. Available from: https://www.migrationpolicy.org/article/ethiopia-origin-refugees-evolving-migration

Adugna, F., Deshingkar, P. and Atnafu, A. (2021) 'Human smuggling from Wollo, Ethiopia to Saudi Arabia: Askoblay criminals or enablers of dreams?', *Public Anthropologist*, 3(1): 32–55.

Al Hammadi, H.M., Bayouomi, K.A. and Bin Rosman, A.A. (2022) 'The impact of abolishing the sponsorship system (kafala) on the peaceful coexistence of population groups in Qatar in light of the migration goals and the alliance of civilizations', *International Journal of Academic Research in Business and Social Sciences*, 12(4): 388–408.

Asnake, A.K. and Fana, G. (2021) 'Introduction', in A. Kefale and F. Gebresenbet (eds) *Youth on the Move: Views from below on Ethiopian International Migration*, London: Hurst & Company, pp 1–20.

Ayalew, M., Aklessa, G. and Laiboni, N. (2019) *Women's Migration on the African-Middle East Corridor: Experiences of Migrant Domestic Workers from Ethiopia*. Available from: https://www.gaatw.org/publications/Ehhiopia_Country_Report.pdf

Ayalew, T.M. (2019) 'The smuggling of migrants from the Horn of Africa through Libya: processes, practices and impacts', in G. Sanchez and L. Achilli (eds) *Critical Insights on Irregular Migration Facilitation: Global Perspectives*, Florence: European University Institute, Robert Schuman Centre for Advanced Studies, pp 6–9.

Beck, D.C., Choi, K.R., Munro-Kramer, M.L. and Lori, J.R. (2017) 'Human trafficking in Ethiopia: a scoping review to identify gaps in service delivery, research, and policy', *Trauma Violence & Abuse*, 18(5): 532–543.

Bird, L. and Reitano, T. (2019) 'Trafficking in persons in conflict contexts: what is a realistic response from Africa?', Policy Brief, ENACT. Available from: https://enactafrica.org/research/policy-briefs/trafficking-in-persons-in-conflict-contexts-what-is-a-realistic-response-from-africa

Carruth, L. and Smith, L. (2022) 'Building one's own house: power and escape for Ethiopian women through international migration', *Journal of Modern African Studies*, 60(1): 85–109.

Chernet, A.G. and Cherie, K.D. (2020) 'Prevalence of intimate partner violence against women and associated factors in Ethiopia', *BMC Women's Health*, 20: 22.

Cockbain, E. and Bowers, K. (2019) 'Human trafficking for sex, labour and domestic servitude: how do key trafficking types compare and what are their predictors?', *Crime, Law and Social Change*, 72: 9–34.

De Regt, M. (2010) 'Ways to come, ways to leave: gender, mobility and illegality among Ethiopian domestic workers in Yemen', *Gender and Society*, 24(2): 237–260.

Demissie, F. (2018) 'Ethiopian female domestic workers in the Middle East and Gulf States: an introduction', *African and Black Diaspora*, 11(1): 1–5.

Eresso, M.Z. (2019) 'Sisters on the move: Ethiopia's gendered labour migration milieu', *Canadian Journal of African Studies/Revue canadienne des études africaines*, 53(1): 27–46.

Estifanos, Y.S. and Freeman, L. (2022) 'Shifts in the trend and nature of migration in the Ethiopia–South Africa migration corridor', *Zanj: The Journal of Critical Global South Studies*, 5(1/2): 59–75.

Fernandez, B. (2011) 'Household help? Ethiopian women domestic workers' labor migration to the Gulf countries', *Asian and Pacific Migration Journal*, 20(3–4): 433–457.

Fernandez, B. (2021) 'Racialised institutional humiliation through the Kafala', *Journal of Ethnic and Migration Studies*, 47(19): 4344–4361.

Gezahegne, K. and Bakewell, O. (2022). *National and International Migration Policy in Ethiopia*. Effext Background Paper, CMI. Available from: https://www.cmi.no/publications/8427-national-and-international-migration-policy-in-ethiopia

Gezie, L.D., Yalew, A.W., Gete, Y.K. and Samkange-Zeeb, F. (2021) 'Exploring factors that contribute to human trafficking in Ethiopia: a socio-ecological perspective', *Globalization and Health*, 17: 76.

Goodey, J. (2008) 'Human trafficking: sketchy data and policy responses', *Criminology and Criminal Justice*, 8(4): 421–442.

Grabska, K. and de Regt, M. (2022) 'Girls on the move: changing dynamics of migration in the Horn of Africa', in J.N Bach (ed.) *Routledge Handbook of the Horn of Africa*, London: Routledge, pp 570–581.

Harmon, R., Arnou, D. and Park, B. (2022) 'TIP for Tal: political bias in human trafficking reporting', *British Journal of Political Science*, 52(1): 445–455.

Henry, A. (2018) *Ethiopian Domestic Workers: Longing for Home*, Pulitzer Center. Available from: https://pulitzercenter.org/stories/ethiopian-domestic-workers-longing-home

Hirschman, A.O. (1981) *Essays in Trespassing: Economics to Politics and Beyond*, Cambridge: Cambridge University Press.

Horwood, C. (2022) 'The war of words in the politization of human smuggling', in C. Horwood, B. Frouws and R. Farin (eds) *Mixed Migration Review 2022: Highlights, Interviews, Essays, Data*, Geneva: Mixed Migration Center.

ILO (International Labour Organization) (2021) 'Enabling access to justice for African domestic workers accused of "absconding" and "theft" in Lebanon'. Available from: https://www.ilo.org/beirut/projects/fairway/WCMS_831955/lang--en/index.htm

IOM (International Organization for Migration) (2021) *They snatched from me my own cry: The interplay of social norms and stigma in relation to human trafficking in Ethiopia*, Case Study: Jimma and Arsi zones. https://publications.iom.int/books/they-snatched-me-my-own-cry

Kefale, A. and Gebresenbet, F. (eds) (2021) *Youth on the Move: Views from below on Ethiopian International Migration*, London: Hurst & Company.

Landau, L.B. (2018) 'A chronotope of containment development: Europe's migrant crisis and africa's reterritorialisation', *Antipode*, 51(1): 169–186.

Longva, A.N. (1997) *Walls Built on Sand: Migration, Exclusion and Society in Kuwait*, Boulder, CO: Westview Press.

Makooi, B. (2020) 'Abandoned by employers, Ethiopian domestic workers are dumped on Lebanon's streets', *France 24*, 25 June. Available from: https://www.arabnews.com/node/2023046/middle-east

Murphy, M., Jones, N., Yadate, W. and Baird, S. (2021) 'Gender norms, violence and adolescence: Exploring how gender norms are associated with experiences of childhood violence among young adolescents in Ethiopia', *Global Public Health*, 16(6): 842–855.

Nisrane, B.L., Ossewaarde, R. and Need, A. (2020) 'The exploitation narratives and coping strategies of Ethiopian women return migrants from the Arabian Gulf', *Gender, Place & Culture*, 27(4): 568–586.

Pucci, P. and Vecchio, G. (2019) 'Trespassing for mobilities: operational directions for addressing mobile lives', *Journal of Transport Geography*, 81.

Reda, A.H. (2018) 'An investigation into the experiences of female victims of trafficking in Ethiopia', *African and Black Diaspora*, 11(1): 87–102.

Robinson, K. (2021) *What Is the Kafala System?* Council of Foreign Relations. Available from: https://www.cfr.org/backgrounder/what-kafala-system

Schewel, K. (2022) 'Aspiring for change: Ethiopian women's labour migration to the Middle East', *Social Forces*, 100(4): 1619–1641.

Semahegn, A. and Mengiste, B. (2015) 'Domestic violence against women and associated factors in Ethiopia: a systematic review', *Reproductive Health*, 12: 1–12.

Shewamene, Z., Zimmerman, C., Hailu, E., Negeri, L., Erulkar, A., Anderson, E., Lo, Y., Jackson, O. and Busza, J. (2022) 'Migrant women's health and safety: why do Ethiopian women choose irregular migration to the Middle East for domestic work?', *International Journal for Environmental Research and Public Health*, 19: 13085.

Sørensen, N.N. (2019) *La vida antes, durante y despúes de la trata: Enfoques innovadores para el estudio de la trata dentro y fuera la República Dominicana. [Life before, during and after Human Trafficking: Innovative Approaches to the Study of Trafficking within and beyond the Dominican Republic]*, Santo Domingo: Instituto Nacional de Migración de la República Dominicana, Ministry of the Interior and Police.

United States Government (2022) *Trafficking in Persons Report: Ethiopia*.

United States Government (2023) *Trafficking in Persons Report: Ethiopia*.

Weitzer, R. (2014) 'New Directions in Research on Human Trafficking', *Annals, AAPSS*, 653: 39–68.

Zaazaa, B. (2022) 'Ethiopian woman is first domestic worker in Lebanon to file slavery case against employer', *Arab News*, 11 February. Available from: https://www.arabnews.com/node/2023046/middle-east

Zeleke, M. (2019) 'Too many winds to consider; which way and when to sail! Ethiopian female transit migrants in Djibouti and the dynamics of their decision-making', *African and Black Diaspora*, 12(1): 49–63.

PART IV

The Issue of Finance

11

Climate-Related Mobility, Land and Inclusive Finance in Rural Ethiopia

Neil Webster and Adane Alemayehu Tadesse

Introduction

What factors lie behind the decisions of households regarding mobility? Is it useful to take the household as the focal point for such decision making? Let us begin by stating that the concept of household is used to denote its importance as an 'economic unit' in decision making. In so doing, it draws upon a rich body of academic literature that addresses decision making in households, based in an agrarian economy, that do not focus solely on commercial marketing of their production, retaining degrees of subsistence farming (Hunt, 2008; Banaji, 2016; Bryceson, 2019). While in many cases the individual's agency within a household can be a major factor, it is proposed here that it is the reproduction of the household as an economic entity as perceived by individual members that shapes the decision making on mobility by household members. This is not to see the household as an individual by another name (Bakewell, 2010); it is to recognize the social embeddedness of individual behaviour and to contest the individualization of mobility (Ruhrort and Allert, 2021).

In the case of Ethiopia, slow-onset climate change in a predominantly agrarian economy characterized by smallholder farming with a strong element of subsistence cultivation is often provided as a causal explanation for mobility, both domestic and international (Belay et al, 2017; Etana et al, 2020a). In this chapter, the proposal is that while slow-onset climate change is an important factor, the basis for the mobility decisions taken by agrarian households varies greatly. Factors influencing these decisions include environmental factors such as the altitude and condition of the land, temperature changes and shifts in rainfall patterns, and changes in degree and nature of vegetation coverage. To these must be added the role of key institutions and not least the presence or absence of government provision in areas such as infrastructure provision, service provision, and the nature and regulation of land rights. The specific characteristics of a household also play a key role. These include the assets it possesses and its constitution along the lines of gender, age, education, employment, social and cultural identity. A factor not so often noted, but possessing considerable influence for the decisions taken is the prevailing discourses on mobility in a locality. These can emanate from international organizations working with migration and mobility, from government organizations seeking to manage such actions, and from social networks used by household members to facilitate and explain their mobility practices (see Moulert et al, 2016; Webster, 2023).

Is there a need for the analytical separation of international migration from other forms of mobility practised by household members? Many in the worlds of research, policy and practice appear to see this as a self-evident need: research centres and departments established to work on migration, international organizations with a primary focus on migrants and migration, governments with policies and programmes directed specifically at migration. However, a key part of the argument in this chapter is that while the forms of mobility involved might possess different names and carry significant differences rooted in legal and political distinctions, from a household perspective they are on the same continuum. The needs that different mobility practices address are not different and the contributions made to a household's livelihood condition do not fall into different categories. In the final reckoning, money is money.[1]

The mobility dynamics

Across the world, mobility in the context of climate change has emerged as an important subject for the research and policy communities. Whether mobility is viewed as a manifestation of crisis or adaptation, it is a commonly practised by households in areas affected by climate change. In the case of Ethiopia, data collected through surveys and interviews for the Governing Climate Mobility (GCM) research programme[2] indicate a considerable degree of continuity in the challenges faced and diversity in the mobility practices pursued. The latter range from immobility to international migration, from short-term and seasonal to long-term relocation, and various patterns of rural–rural and of rural–urban movement. The many forms of multiple mobility practised are very much shaped by the capabilities and socioeconomic position of households. For instance, the most vulnerable households are often restricted to short-distance mobility, whereas those with relatively better capabilities move further afield and tend to benefit the most.

Why practise mobility? Even in the context of climate change, mobility is not just driven by climate change. The explanations offered by respondents indicate a range of influencing factors. These include the distance and time taken to reach markets, the possibility of alternative employment, the unpredictability of rains, the soil characteristics of their land, the availability and quality of education and health in their present locality, whether an access road is asphalt or not, their access to electricity and clean drinking water, and their general vulnerability to poverty.[3] Rural households in the study areas indicate that slow-onset climate change hazards have resulted in the depletion of household assets and livelihood crises, and that these have had a direct influence in their consideration of mobility practices as an option.

Mediating how such factors are experienced at the individual level are factors of age, gender, education level, health condition, own wealth (or debts), religious and sociocultural identity, familial and social networks that the individual carries, and through which she is perceived by others. Analytically, these characteristics can be approached in terms of diverse sets of capabilities and aspirations possessed, which in turn shape the agency present (Appadurai, 2004; de Haas, 2014; Ruhrort and Allert, 2021). Most importantly, finance has emerged as a central factor in shaping both the aspirations and capabilities of members of rural households with regard to their adoption of mobility or immobility.

The importance of access to finance

Access to finance is a powerful factor in shaping the plans of households, not least in setting the aspirational horizons as to what is perceived to be possible (Appadurai, 2004; de Haas, 2014). As such, it must be viewed as important in shaping the agency practised. Finance is central to a household's capability to pursue specific goals and, through these, make progress towards a broader set of family and household objectives. In the absence of finance possessed, it requires credit. Credit varies greatly in terms of its form and its accessibility. Access to a source of institutionally regulated credit is preferable, though not always desirable, on the part of rural households in less developed countries. Institutions can be remote and inflexible; family and friends can be more accommodating; institutional credit is often found in urban areas and far from many rural localities; and often there is an inbuilt bias against lending to small farmers as they are perceived as having poor collateral and poor 'credit discipline'. Berg and Emran (2020) note that institutionalized micro-credit arrangements in rural Bangladesh tend to neglect those with no or smaller-sized farmlands. All too often, the small landholding farmer must turn to more informal sources of credit in their locality. Private moneylenders take high rates of interest, are known to take repayments in farm outputs valued at very low market rates just after harvesting, and such loans can be linked to securing strong relations of dependency between lender and recipient. When accessible, the formal banking sector can be an avenue to credit that is more regulated, more predictable, carries fewer social and political obligations, and is generally more favourable in its terms – for example, the rate of interest is lower, the contractual terms clearer and they carry fewer 'hidden' ties in the forms of payments.[4]

Etana et al (2020b) also point to the tendency for households studied in Ethiopia to develop fatalistic attitudes and risk-averse behaviour when faced with long-term economic problems. This can result in the non-use of adaptation strategies over time, resulting in a form of path dependency that requires effective institutional interventions if it is to be changed. Government-led inclusive finance is suggested here as one such intervention, potentially a very powerful one.

In the 1990s and 2000s, micro-finance was the promoted with considerable fervour by bilateral and multilateral donors and a great number of nongovernmental organizations (NGOs), both international and national. More recently, inclusive finance has been adopted as the aid community's favoured terminology. Importantly, it comes with a healthy concern over the objectives seen to be achievable with inclusive finance activities insofar as they might go beyond the purely economic. But there remains considerable debate as to whether it can promote development and not least more equitable development. This places an important question as to the degree that economically poor households can benefit in a meaningful way from greater access to financial services. There is also the matter of involving the private sector in such programmes, and banks in particular, that may evaluate whether financial inclusion makes good business sense (Mader, 2017).

With access to institutional credit provided through a formal institutional framework, standardized and regulated by a government representing the interests of an increasingly broader set of its citizenry (polity), not only is economic development seen to occur, but equality in the distribution of the gains is also achieved through production and redistribution (Pazarbasioglu et al, 2020; UNCDF, 2021). However, both these are argued here to be very much defined by the nature of the political regime, its approach towards inclusive finance, that are in turn underwritten by the social and political interests it serves. In an ideal situation:

> An inclusive financial market provides affordable and equitable access to financial products for all households and entrepreneurs, especially the most marginalized. By empowering people to exploit a wider set of economic opportunities, inclusive finance can therefore be a pivotal tool in driving economies on a sustainable growth strategy. (Corrad and Corrado, 2017: 19)

What is perhaps understated is that inclusive finance requires not just banking services being available to all within a locality, but that these are also accessible to all. In this, affordability is just one element; the socioeconomic positions that households occupy are often more significant determinants of financial inclusion or exclusion. So, setting affordability to one side for a moment, financial inclusion requires a major shift in existing social and political relations. Such a shift will be opposed by some if not many. From a purely economic perspective, the question is as to whether the recipient of financial services can pay for them. From a social perspective, the question is as to whether it is desirable that a particular individual becomes a recipient. From the individual's perspective, do they expect to be heard inside the doors of such an institution? The latter two questions concern normative values, deep-seated attitudes and, not least, what we would term as 'hierarchies of exploitation'.[5]

Not just the lending of money

Lack of access to financial services can be a factor in preventing poorer households from adapting to climate change in Ethiopia: the purchase of new agricultural inputs, investments in irrigation, enhancing capacities to physically access markets by way of transportation and much more. As indicated earlier, for most agrarian households, this requires a degree of credit, initially short term and possibly seasonal, but usually with a longer time horizon. Mobility practices also require an ability to transfer funds.

The last two decades or so have seen quite dramatic changes in this area as banks have developed a growing infrastructure enabling funds to be moved from region to region within a country as well as internationally from one country to another. The availability of formal banking services has grown exponentially in many countries, not least due to the transfer of remittances from workers to their families. But this is not necessarily the case for all households having members working elsewhere. Relatively lower accessibility to banking services for poorer households serves to reinforce existing inequalities both in the countryside and in urban and semi-urban areas.

From the data collected in the two GCM field sites in Ethiopia, making finance more inclusive appears to lie in the availability and the accessibility of the formal banking sector. As indicated earlier, the former addresses the physical or virtual presence of banks. The latter addresses the socioeconomic and political status of a household as seen by others within the social setting as well as by the financial institutions. To this must be added a household's self-perception of its social and economic standing in relation to a financial institution such as a bank in terms of their perceived 'social eligibility' for the financial services it might offer.[6]

The field sites are in West Arsi zone of the Oromia region and in the South Wollo zone of the Amhara region. In general terms, it can be said that the three kebeles in West Arsi are close to a significant urban centre, are low-lying in comparison to South Wollo, and have relatively fewer hills and steep valleys. South Wollo, while close to a market town, is more rural in its agrarian characteristics. The three kebeles are also located at different altitudes on the side of the rift valley and thereby carry some variation in agricultural conditions.

Table 11.1 presents the findings from the 808 households surveyed in the two field-site localities in 2021.[7] As the data show, most of these households had experienced problems in the previous five years, though considerably more so in West Arsi than in South Wollo.

In West Arsi, the direct effects from slow-onset climate change are seen in the amount and pattern of rainfall, temperature increases, and the increase in crop diseases and pests. To these should be added problems linked to poor service provision and poor access to inputs, such as fertilizers and seeds, and poor access to land and productive assets more generally (Rahmato et al, 2021).

In South Wollo, the picture is rather different, the more significant challenges being less related to the direct effects of slow-onset climate change and more to the poor access to inputs, lack of employment and the high cost of food – these problems were also present in West Arsi, but were highlighted less by those interviewed than the direct effects of slow-onset climate change.[8]

For the individual household, the challenges are considerable. A 40-year-old father to five children in Faji Solee kebele, West Arsi, described the challenges he faces:

> During the past five years I am observing a change in the rainfall pattern. Particularly the last three years are very difficult for me and my family. Because we were affected by crop damage caused by erratic rainfall. As a result, I was forced to sell my cattle so that I can cover the cost of living which was supposed to be covered by the income that should have been generated from my farming activities. During the first year when my teff crop was damaged, I did not feel the burden because I had some savings. But during the second and the third year it became too much to handle as my loss was not only the crops but the money I invested for fertilizer and seeds was also gone. (Faji Solee kebele, Shashemene, 29 March 2022)

Is there help to be found from the government side? Table 11.1 already suggests that there are issues with the presence and role of government, especially in West Arsi. As the source of the data is a survey, it is important

Table 11.1: Percentage of households interviewed experiencing a hazard within the last five years

Type of hazard experienced	West Arsi	South Wollo	All
Drought (shortage of rainfall)	83.7	28.2	66.1
Floods	38.8	27.0	52.9
Rains irregular, insufficient or too heavy	91.2	35.0	73.6
Extreme temperatures (high and/or low)	95.6	48.5	72.6
Crop pests/diseases (eg locusts, rodents, birds, fungi and so on)	84.1	67.2	56.2
Decreasing soil fertility	85.9	55.9	64.5
Problems with input purchase (seeds, fertilizers, raw materials and so on)	91.0	49.5	63.9
Problems with output sales (agricultural, artisan, fish, dairy products and so on)	64.1	28.2	41.8
Livestock diseases	58.8	43.1	76.5
Lack of drinking water	86.3	36.8	76.2
Illness affecting the household/family	48.3	19.4	41.0
High food prices	98.8	94.4	60.4
Lack of employment	97.3	65.9	59.0
Problems with access to farming land	89.5	33.6	55.1
Problems accessing forestry and non-timber forest products	77.6	21.8	54.9
Conflicts over land (land grabbing)	66.3	20.6	47.4
Conflicts over service provision (schools, clinics, drinking water, electricity and so on)	57.8	20.6	42.4
Lack of government support to help with households' problems	81.7	32.1	58.5

Source: GCM Household Survey 2021

to recognize that the evidence is based on heads of households' perceptions and experiences; strong secondary data are not available. What is indicated in Table 11.2 is that very little help is available, and a marked lack of government support was reported by a large majority of respondents.

Similarly, key informant interviews conducted in West Arsi also concur with the survey data. For instance, one of the key informants from Haleche Harabate stated that 'there is no financial support from government organizations, NGOs or community-based organisations (CBOs) for the community to help as part of a coping strategy for climate

Table 11.2: Surveyed households' receipt of government assistance in response to drought, flooding, poor/failed harvests or lack of food, % of households

Kebele	Received	Not received	Do not know
Bededo (South Wollo)	8.1	91.9	0.0
Hitecha (South Wollo)	11.8	86.0	2.2
Seglen (South Wollo)	21.3	78.7	0.0
Faji Sole (West Arsi)	4.0	93.5	2.4
Haleche Harabate (West Arsi)	5.7	93.6	0.7
O'ene Chefa Umbure (West Arsi)	15.1	84.9	0.0

Source: GCM Household Survey 2021

change. The informal credit obtained from relatives is the only financial source used to cope in the dry season'. A female head of household in the same locality said that in comparison to wealthy households, the poor are more affected due to limited assets, such as a lack of any stored crops or savings in a bank that could be used in some other activity. She argued that the 'rich' have the capital to engage in trade and similar activities when crop production fails, as had occurred last year. There is no financial service provider of credit, no donor. She believes that if access to credit is there, it could help farming households cope with the adverse effects of climate change.

Providing access to finance is to provide support that can facilitate the agency of the household in taking livelihood decisions. It is an area in which government can have a very significant role, not least in securing access to finance. Such access can help reduce inequality, especially when it reaches those traditionally excluded and when greater access to finance promotes increases in production and employment opportunities. This is a position strongly backed by the World Bank, which argues that 'access to affordable financial services is critical for poverty reduction and economic growth' (Pazarbasioglu et al, 2020: v). Among the several forms of financial services that the Bank examined, remittances and government-to-person payments are seen as particularly beneficial to the poor. By 2021, remittances at the global level were estimated to exceed US$600 billion, an amount greater than foreign direct investment (FDI) and overseas development assistance (ODA) combined (Pazarbasioglu et al, 2020: v).

As previously noted, the increased role of banks in the remittances has seen a digitalization of transfers and a halving of the transaction cost from an estimated 6.8 per cent to 3.3 per cent. However, the economic effects of COVID-19 have been particularly noticeable in terms of their impact

Table 11.3: Accessing of credit by households, % of households interviewed

	West Arsi			South Wollo		
	Regularly	Occasionally	Never	Regularly	Occasionally	Never
Bank	0	1.7	98.3	0	21.1	78.9
Cooperative	1.5	17.8	80.7	0	81.1	18.9
Moneylender	0	18.3	81.7	0	82.8	17.2
Family	0.2	61.7	38.1	0.5	41.7	57.8
Friends	0	65.1	34.9	0.2	44.9	54.9

Source: GCM Household Survey 2021

Figure 11.1: Proportion of households with a bank account by village

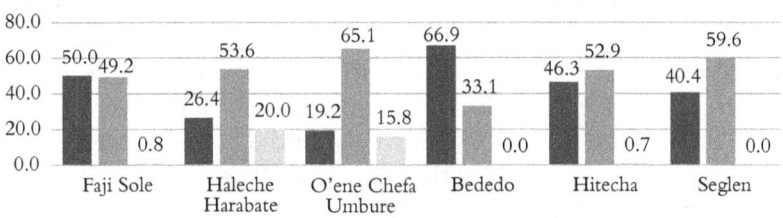

Source: GCM Household Survey 2021

on the level of remittances. One result is that government payments in the form of cash transfers and transfers in kind such as food have taken on a greater role in the search for ways to reduce poverty.

Table 11.3 points to a significant majority of the households interviewed in both field sites as having never taken credit from banks. They might possess a bank account (see Figure 11.1), but it is not a basis for taking loans from a bank. This mirrors a more general finding reported by the International Finance Cooperation (IFC), namely that only 1 per cent of total bank lending in Africa goes to the agricultural sector, despite approximately 55 per cent of the continent's population being employed in agriculture (IFC, 2014).

The fact that households have bank accounts but little in the way of loans can be an important factor in decisions on mobility. It is on this point that the nature of land rights in Ethiopia is important and the extent to which a market in land exists in practice (Ali et al, 2011). The possibility of land acting as loan collateral and the possibility of accumulating land for cultivation were prominent points in interviews and discussions held, and will be taken up later in this chapter.

Social transfers in cash and kind

What of other possible sources of finance, and social benefits in particular, that might be provided by the state? Pankhurst and Dom (2019) suggest that until recently in Ethiopia, social protection was primarily found within communities and there was little external involvement, whether by the state or by other agencies. Policies did exist, such as the Social Protection Policy of 2014, but implementation was slow due to human and financial resource constraints. External responses to famines in the 1970s and 1980s laid the basis for the Productive Safety Net Programme (PSNP) launched in 2005. This receives some government support, but remains heavily donor-dependent for its finances. In addition, significant sections of the population remain outside of this programme as their localities are not defined as being drought-prone.

In West Arsi, no variable was found to be a significant factor in determining whether such a social benefit was provided to a household. In South Wollo, age and land size were found to be significant factors; the presence of elderly family members and of smaller landholdings increasing the probability of a social benefit being received by a member of the household. In addition, having nonresident members in the household also proved significant. The factor of age can be linked to the provision of a state pension, and the size of landholding to food insecurity and the fact that the PSNP is present in this locality, while this is not the case in West Arsi. The latter is not listed as being drought-prone. The significance of nonresident household members in South Wollo is not so easily explained. As the significance is borderline ($P = 0.048$, that is, just under 0.05), it could be a case of correlation rather than causation, but more detailed analysis needs to be undertaken.

Land and the importance of land rights

The importance of land rights and their implications for the way in which land markets work cannot be overstated. In rural areas, households' experiences of insecurity are critical to their decision making. The degree and security of access today and its predictability for the future are a constant consideration. Not only do these shape decisions on what to cultivate, they also shape the decisions of others; security rooted in the ability to sell land or sell the lease of land determine which type of creditor can be approached and the terms of any loans obtained (Ali et al, 2011)

A few argue that investment precedes security in that it can enhance the security of property rights. The evidence is mixed; much depends on the specific nature of land rights possessed. In Ethiopia, the state has had and continues to retain a dominant position on land rights as it formally owns all land and rejects its sale and the exchange of leases (Rahmato, 2004; Ali et al,

2011). Regional governments have continued to distribute and redistribute land after the 1995 Constitution, raising the fear among farmers that they might be subjected to possible land redistribution without compensation. In the two localities studied, investments in the cultivation of chat, eucalyptus and coffee were present. All can be characterized as more permanent investments than crops planted and harvested in the same year. As such, an expectation of a degree of continuity with a farming household's property rights is an important variable. As indicated earlier, this will also play a role in accessing financial credit for the investment, the possible source and terms of such credit and, not least, the decision of the farmer as to whether to invest.

Similarly, the pursuit of remittances as a possible source of investment funds will also be influenced by the perception of security of one's land rights. It will also be influenced by the need for an initial investment required for most mobility practices, particularly those involving residence in urban centres or movement abroad in search of employment. An interesting tendency noted in several interviews is the investment of remittances in urban land, possibly indicating a perceived lack of agricultural land security combined with a desire to access the benefits and opportunities associated with urban areas. Here perceptions tend to focus on better employment and small investment opportunities combining with better access to education, health and other services.

Table 11.4 does point to households in South Wollo having members migrating for longer periods and with a significantly higher number migrating overseas than is the case in West Arsi. While such mobility practice can be interpreted as being a possible adaptation strategy, it can also have implications with respect to the household in South Wollo. The evidence suggests that the most vulnerable tend to be immobile, but other households are turning to mobility, possibly out of a perceived increase in their vulnerability. Interviews suggest this to be the case and household eligibility for the PSNP and similar social benefit payments supports this. However, migration is not necessarily the solution. Interviews point to increased vulnerability when remittances fail to materialize, are linked to illicit/informal migration practices or are simply 'misused' by the recipient family members.

The data in Table 11.4 on households' responses to contemporary challenges can be grouped into three general categories: those based on changing agricultural practices; those based on some degree of movement into nonfarming activities, whether in the form of alternative employment, trade in agricultural products or handicraft production; and those involving mobility practices. The categories are not mutually exclusive – there can well be an overlap between them as with alternative nonfarming employment and migration – but common to all three categories is the pursuance of alternative income to the household economy.

Table 11.4: Proportion of surveyed households adopting a particular strategy by region

Adaptation strategy for household	West Arsi (%)	South Wollo (%)	P-value	Total (%)
Improved irrigation	4.9	30	0.000	17.45
Farming on less land	37.3	38.4	0.732	37.85
Maintaining fewer livestock	35.4	42.9	0.027	39.15
Adopting improved or new crop varieties	50.2	45.9	0.233	48.05
Change time of planting	49	32.5	0.000	40.75
Use more fertilizer and pesticide	42.2	52.9	0.003	47.55
Use more machinery	37.3	3.6	0.000	20.45
Take steps against soil erosion	35.6	54.3	0.000	44.95
Buy crop/cattle insurance	0.5	17.6	0.000	9.05
Seasonal migration by household members to urban areas	37.1	30	0.030	33.55
Seasonal migration by household members to rural areas	29.3	36.4	0.033	32.85
Permanent migration household members (minimum five years)	10.5	33.1	0.000	21.8
Short- or long-term migration abroad	9.8	29.1	0.000	19.45
Start to produce handicrafts	15.1	17.6	0.426	16.35
Look for non-agricultural work	27.3	45.7	0.000	36.5
Begin trade in agricultural and other commodities	19.8	50.1	0.000	34.95
Reduce expenses (for example, type and number of meals)	52.4	47.1	0.142	49.75

Source: GCM Household Survey 2021

Land and loans

Taking a step back, what is the situation regarding access to micro-finance? Again, the conflict directly affecting the northern field site has resulted in the focus here being on West Arsi. Though both government and private micro-finance institutions are operating in the zone, their services are not accessible to all. The major challenge in this regard is strict eligibility criteria that give more room for existing assets and social capital. In most cases, poor rural households are not able to secure credit from microfinance institutions.

Despite the existence of several micro-finance institutions in West Arsi, mainly Oromia Credit and Savings Share Company (OCSSC) and Buusaa

Gonofaa Micro-Finance Share Company (BGMFSC) are operating in the study sites. Here it is worth noting that besides socioeconomic positioning of households, the rural-urban divide is another important factor that shapes access to financial services, whereby the rural is significantly marginalized.

Oromia Credit and Saving SC (OCSSC) is a state affiliated micro-finance institution that has been providing credit to the rural communities in the Oromia region, including West Arsi, for more than two-and-a-half decades. For a long period of time, though other private and NGO-affiliated micro-finance institutions like Buusaa Gonofaa micro-finance SC are operating in the area, OCSSC remained the primary micro-finance service provider in rural West Arsi. In 2021, OCSSC evolved into Sinqe Bank (Baankii Siinqee). Though there are changes in certain areas like increasing highest loan limit from 15,000 birr to 275,000 birr, the organization has retained its main role in providing rural micro-finance services.[9]

The organization offers three type of loans/credit, namely the Agriculture Based Individual Loan (ABIL), the group loan and the cooperative loan. Rural households in Shahsemene are eligible for ABIL and group loans. ABIL is a credit system that targets on financing agricultural engagements of ordinary rural households. Though each eligible household is entitled to take a loan of 15,000 birr, it is unlikely that households will be able to take the maximum amount as their request is subject to evaluation of a customer relations officers and a committee that comprises 'model farmers' and local government officials. In connection with eligibility criteria, in most cases 'model farmers'[10] are the primary beneficiaries of ABIL. Here it should be noted that access to ABIL is highly influenced by the model farmers community as they participate in screening and verification process through the committee structure. On the other hand, the eligibility criteria perfectly fit the profile of model farmers. Moreover, model farmers' unreserved access to local government offices also gives them an extended access to the credit system. Eligibility to ABIL is evaluated based on previous credit history, asset, endorsement from the local credit committee, confirmation of permanent residency, land certificate or book, and clearance from other micro-finance institutions.[11] Nevertheless, the loan can only be issued if the applicant is able to bring urban house ownership certificate, a vehicle ownership certificate or a letter of guarantee for government employment. Farmland cannot be considered as collateral. In the case of the group loan, rural households are expected to forge a group of ten or more individuals who want to take credit from OCSSC. Rural households are not required to present collateral or a letter of guarantee. As a collective loan, members of the group provide guarantees for each other. The maximum limit to this type of loan was 30,000 birr. Cooperative loans, on the other hand, were introduced to make funding available for those who organized themselves as small and medium-sized enterprises. As a result, rural households were not eligible, whereas individual members that formed small and medium-sized enterprises were eligible.

During the 2021/2022 fiscal year, only six rural households were successful in taking a loan from the ABIL programme, while the group loan was taken by 31 groups constituting 257 people or households. In the case of ABIL the maximum limit for a single household is 275,000 birr, those six households that took this loan in the 2021/2022 in full were each given 200,000 birr.[12] When compared with previous years, the number of people taking loans has declined due to procedural and religious factors. Procedurally, OCSSC's attempt to comply with the National Bank of Ethiopia's provision that demands a 97 per cent recorded loan return is diminishing the number of kebeles that are covered by OCSSC. In particular, kebeles that are frequently affected by drought and crop failure are being excluded from engaging with micro-finance institutions. There is also the role played by religion as the majority in West Arsi are Muslims and many do not wish to take a loan on which interest is paid for religious reasons.

Rural households are eligible for credits if they become a member of OCSSC's saving scheme and save funds for at least six months. However, eligibility for credit requires an endorsement from the local loan officer and from a committee established by the local government unit.[13] It was found that the committee would not issue a letter of guarantee to rural households that have a track record of leasing their land and those with less than 0.5 hectares of land.[14] In this way, it can be seen that access to credit from OCSSC is mediated by a combination of organizational procedures and local actors, who consider the socioeconomic and political positionality of individual households. Perhaps unsurprisingly, households that are ranked as rich or model farmers have easier access to a loan. A female participant in a focus group discussion stated that: 'Getting a loan from OCSSC was not that much difficult. I was required to save money for some time and then they gave me the money [loan] by taking only the copy of the land certificate of my farmland as collateral for the loan.'[15]

This response again indicates that land is a necessary asset for facilitating access to credit, even though using farmland as collateral is in fact technically illegal. Others indicated that they were denied loans because they did not own a house in the town.[16] This suggests that access to credit from micro-finance institutions is tilted towards 'rich households' with favourable interpretations of their land rights, namely regarding what is formally a lease right as equivalent to an ownership right. Meanwhile 'poor households' are subjected to lengthy informal procedures that involve the local government officials' somewhat subjective assessment of their economic condition and not least the status of their land rights.

Members of rural community in West Arsi, who are unable to access credit from micro-finance institutions, are forced to enter into more exploitative financial relationships. Poor and middle-income households seeking loans out of necessity stated that they turn to private money lenders in the local business community. These demand high interest rates for providing credit, but as they do not have any strict eligibility criteria, they are more accessible. While this source of financial credit is viewed as a remedy for short-term financial problems due to the high level of interest taken, the negative impact on recipient households is considerable. Many must pay an annual interest that can reach 100 per cent of the original loan. In the worst-case scenario, the inability to repay the loan and interest results in the loss of land and other assets used as collateral at the time of taking the loan. Participants in focus group discussions stated that private money lenders tended to target a poor household's land when making these informal credit agreements.

Poor rural households were also found to be entering into land transactions that involved renting or selling their farmland to 'model' farmers thereby increasing the inequality present in rural communities. The research shows quite clearly that economic pressures rooted in the effects of climate change-related risk are forcing the poor households to enter into unfavourable land transaction arrangements in the absence of alternatives. In so doing, the existing policy on land rights in which the local government controls land transaction activity is being circumvented. This policy places restrictions on the eligibility to rent land and sets the period of the contract. In practice, rural households are moving towards a dual-track approach involving two different contracts – one as per the law and the other based on the actual terms of the agreement reached.

The potential for digital inclusive finance

As noted by Montfaucon (2020), Karakara and Osabuohien (2019) and the World Bank (2016), financial transactions are increasingly digital in contemporary Africa and based on mobile telephones and to some extent the internet. In all three kebeles in the South Wollo field site, over 82 per cent of surveyed households possessed at least one mobile phone. However, only 3.7 per cent had access to the internet in one way or another. In the West Arsi field site, the surveyed household ownership of mobile phones was just over 80 per cent and access to the internet was 2.4 per cent.[17]

The prevalence of mobile phones and the increasingly common practice of transferring funds across distances by family members point to the considerable potential in Ethiopia linked to digital financial services (Rapsomankis, 2015). It is a financial infrastructure that does not need the same investment that more formal banking services would require if they were to be accessible to the rural majority. Remittances are already playing an influential role, in that these have introduced new ways and means for transferring funds across considerable distances. Many households can now access these with the use of their mobile phones. They bring former non-users into the digital room of finance. Social and cultural factors that have often reproduced exclusionary practices are pressed to one side as monies are increasingly sent through banks rather than through the informal agents traditionally used for this purpose. Not only are they quicker, cheaper and offer greater control to the user, but informal financial sources also tend to reproduce and promote inequality due to the higher costs of borrowing and transfers as well the social dependencies often involved.

There is a strong and compelling argument today that a strategy aimed at securing more inclusive finance should challenge the 'hierarchies of exploitation'. These enable those with control over key resources such as land, inputs, transportation, markets and so on to maximize their returns by providing finance, whether in kind or cash, to those locked in dependency relations centring around land, production and markets for inputs and outputs.

Conclusion

Here it is instructive to consider the capacities for transformative and transactional change that strategies for inclusive finance possess. It is a distinction that is important to understand in order to note changes that might be achieved, but also changes that might be prevented from occurring. Transformative change would involve changes in the way in which the government engages with the citizens, allocates its funds, and provides its services and resources. It would involve the government seeking to facilitate the agency of all rural households through the ways that finance is made accessible. Transactional changes can be measured in more quantitative ways, such as the number of days that household members work, whether it is on the landholding or elsewhere, and the number of cattle owned. These are absolute changes and they certainly determine the household's livelihood condition and its wellbeing. However, relational changes that are transformative are not only absolute in nature, they can also ensure that absolute gains are more sustainable, are intergenerational and can be defended legally as well as politically or morally. Failure to embrace this analytical distinction might well result in a failure to build on both when implementing activities aimed at more inclusive finance.

Are the approaches of institutions promoting inclusive finance too technocratic to bring about transformative changes through inclusive finance? Do they unwittingly reinforce the political settlements[18] that perpetuate such relations and the inequalities they sustain? At best, one can say that the judgement cannot be made as the activities have yet to be rolled out. Yet it is possible to conclude that, at this point in time, the insecurity associated with access to land is significant in hindering any inclusive finance strategy achieving its aims.

To conclude with an example that illustrates the challenge and not least the potential bias in economic development in rural Ethiopia: coffee, eucalyptus and khat are three commercial crops found in localities covered by the research programme. Coffee begins to yield a commercially viable crop after three to four years, but only reaches its full potential after eight years or so. Eucalyptus and khat can yield substantial returns after two to three years. If coffee is a longer-term investment with many years of high production, eucalyptus and khat are investments with a medium-term horizon. The potential for many more farmers, not least smallholder farmers, to engage in commercial agriculture is considerable. The risks in doing so also remain considerable. Enabling more accessible finance could shape the space in which decisions on mobility are being taken. Mobility would undoubtedly remain with a successful inclusive finance strategy, but its form and role could be changed significantly, not least contributing to a more transformative approach in Ethiopia's agrarian economy.

Notes

[1] See Olwig and Sørensen (2002: 7) for a similar argument.
[2] GCM was a collaboration between the Centre for Migration Studies at the University of Ghana, the Forum for Social Studies in Ethiopia, and the Danish Institute for International Studies in Denmark. The research was provided with funds from the Danish Ministry of Foreign Affairs (2019–2023).
[3] Vulnerability is rooted in many factors, including illness in the family, insecure land tenure, drought and irregular rains, indebtedness, age and poor access to markets.
[4] Hidden costs to the loan taker can include payments in kind with very unfavourable market prices applied, and payments requiring the provision of labour at lower than market rates or even unpaid, as in the case of domestic work provided by a family member.
[5] Social, cultural and political hierarchies are closely linked to ways of extracting and accruing capital (economic, political and social), but all at some point assuming an economic form.
[6] Those without financial services are more likely to be less educated, and almost a fifth cite distrust as a reason to refrain from using financial services (Demirgüc-Kunt et al, 2018).
[7] It should be noted that the survey was undertaken prior to the conflict in Tigray.
[8] See Rahmato, Mohammed and Webster (2021) for a detailed description of the two localities.
[9] Interview with Branch Manager of Sinqe Bank, Shashemene, 1 April 2022.
[10] Model farmers are understood to be farmers who display a higher level of productivity due to their 'best practices', which they are expected to disseminate to other farmers through peer-to-peer learning in farmer networks (FDRE, 2016; MoANR and ATA, 2017). This strategy is expected to increase Agricultural Extension Service coverage and improve technology use (MoANR and ATA, 2017). Similarly, as agriculture becomes more commercialized, model farmers are expected to play a strategic role in organizing farmers and in serving as role models in relation to produce for the market (Davis et al, 2010). Berhanu and Amdework (2011) stated that the understanding of 'model farmers' on the ground is broader than what is anticipated in government documents. At the local level, a model farmer is 'not only expected to exhibit exemplary farm practices

(for example, proper management of soils, correct application of chemical fertilizers and seeds) but also engage in activities which are considered innovative and new in the area' (Berhanu and Amdework, 2011: 14). In this regard, model farmers are also viewed as 'entrepreneurs' who are extensively engaged in nonfarming business activities (Berhanu and Amdework, 2011) that in turn have enabled them to build an asset and economic capacity. Lefort (2012) argues that in the post-2005 Ethiopia, 'model farmers' are identified as allies of the ruling party who are expected to facilitate the hegemony of the party, while the regime vowed to 'open the way to their economic prosperity'. Therefore, model farmers could be taken as local elites who dominate the political and economic sphere of the rural Ethiopia.

[11] Interview with Branch Manager of Sinqe Bank, Shashemene, 1 April 2022.
[12] Interview with Branch Manager of Sinqe Bank, Shashemene, 1 April 2022.
[13] Interview with Kebele Manager of O'ene Chefa Umbure Kebele Administration (Shashemene), 30 March 2022.
[14] Interview with Kebele Manager of O'ene Chefa Umbure Kebele Administration (Shashemene), 30 March 2022.
[15] Focus group discussion with local community members, O'ene Chefa Umbure Kebele (Shashemene), 25 March 2022.
[16] Focus group discussion with local community members, O'ene Chefa Umbure Kebele (Shashemene), 25 March 2022.
[17] Perhaps noteworthy is that while over 50 per cent of mobile phones were described as jointly owned by husband and wife, where they were individually owned, then it was usually the man who possessed a mobile phone, unless it was a female-headed household.
[18] Political settlement is used to denote the presence of elite coalitions designed to secure their collective interests (Khan, 2018).

References

Ali, D., Dercon, S. and Gautam, M. (2011) 'Property rights in a very poor country: tenure insecurity and investment in Ethiopia', *Agricultural Economics*, 42: 75–86.

Appadurai, A. (2004) 'The capacity to aspire: culture and the terms of recognition', in V. Rao and M. Walton (eds) *Culture and Public Action*, Stanford: Stanford University Press, pp 59–84.

Bakewell, O. (2010) 'Some reflections on structure and agency in migration theory', *Journal of Ethnic and Migration Studies*, 36(10): 1689–1708.

Banaji, J. (2016) 'Merchant capitalism, peasant households and industrial capitalism: integration of a model', *Journal of Agrarian Change*, 16(3): 410–431.

Belay, A., Recha, J., Woldeamanuel, T. and John, F. (2017) 'Smallholder farmers' adaptation to climate change and determinants of their adaptation decisions in the Central Rift Valley of Ethiopia', *Agriculture & Food Security*, 6(24):1–13.

Berg, C. and Emran, M. (2020) 'Microfinance and vulnerability to seasonal famine in a rural economy: evidence from Monga in Bangladesh', *Journal of Economic Analysis & Policy*, 20(3): 1–36.

Berhanu, A. and Amdework, E. (2011) *Peasant Entrepreneurship and Rural Poverty Reduction: The Case of Model Farmers in Bure Woreda, West Gojjam Zone.* Addis Ababa, Forum for Social Studies.

Bryceson, D. (2019) 'Gender and generational patterns of African deagrarianization: evolving labour and land allocation in smallholder peasant household farming, 1980–2015', *World Development*, 113(2019): 60–72.

Corrado, C. and Corrado, L. (2017) 'Inclusive finance for growth and development', *Current Opinion in Environmental Sustainability*, 24: 19–23.

Davis, K., Swanson, B., Amudavi, D., Daniel, M., Flohrs, A., Riese, J., Lamb, C. and Zerfu, E. (2010) 'In-depth assessment of the public agricultural extension system of Ethiopia and recommendations for improvement', *International Food Policy Research Institute Discussion Paper*, 1041: 193–201.

De Haas, H. (2014) 'Migration theory: quo vadis', *Zenodo*. DOI 10.5281/zenodo.12681.

Demigüc-Kunt, A., Klapper, L. and Singer, D. (2018) *Financial Inclusion and Inclusive Growth A Review of Recent Empirical Evidence*, World Bank Policy Research Working Paper 8040.

Etana, D., Snelder, D.J., van Wesenbeeck, C.F. and de Cock Burning, T. (2020a) 'Trends of climate change and variability in three agro-ecological settings in central Ethiopia: contrasts of meteorological data and farmers' perceptions', *Climate*, 8(121): 1–27.

Etana, D., Snelder, D.J., van Wesenbeeck, C.F. and de Cock Burning, T. (2020b) 'Dynamics of smallholders' adaptation decision-making in Central Ethiopia', *Sustainability*, 12(11): 1–26.

FDRE (Federal Democratic Republic of Ethiopia) (2016) *Growth and Transformation Plan II (GTP II) (2015/16−2019/20)*. Available from: https://ethiopia.un.org/en/15231-growth-and-transformation-plan-ii

Hunt, D. (2008) 'Chayanov's model of peasant household allocation', *Journal of Peasant Studies*, 6(3): 247–285. DOI: 10.1080/03066157908438075

IFC (International Finance Corporation) (2014) *Access to Finance for Smallholder Farmers*. Washington DC: International Finance Corporation/World Bank Group.

Karakara, A. and Osabuohien, E. (2019) 'Households' ICT access and bank patronage in West Africa: empirical insights from Burkina Faso and Ghana', *Technology in Society*, 56: 116–125.

Khan, M. (2018) 'Political settlements and the analysis of institutions', *African Affairs*, 117(469): 636–655.

Lefort, R. (2012) 'Free market economy, "developmental state" and party-state hegemony in Ethiopia: the case of the 'model farmers', *Journal of Modern African Studies*, 50(4): 681–706.

Mader, P. (2017) 'Contesting financial inclusion debate', *Development and Change*, 49(2): 461–483.

MoANR (Ministry of Agriculture and Natural Resources) and Agricultural Transformation Agency (ATA) (2017) *Ethiopia's Agricultural Extension System: Vision, Systemic Bottlenecks and Priority Interventions*, Addis Ababa: Federal Democratic Republic of Ethiopia, Ministry of Agriculture.

Montfaucon, A. (2020) 'Increasing agricultural income and access to financial services through mobile technology in Africa: evidence from Malawi', in E. Osabuohien (ed.) *The Palgrave Handbook of Agricultural and Rural Development in Africa*, London: Palgrave Macmillan, pp 247–262.

Moulaert, F., Jessop, B. and Mehmood, A. (2016) 'Agency, structure, institutions, discourse (ASID) in urban and regional development', *International Journal of Urban Sciences*, 20(2): 167–187.

Olwig, K. and Sørensen, N. (2002) 'Mobile livelihoods: making a living in the world', in *Work and Migration: Life and Livelihoods in a Globalizing World*, London: Routledge, pp 1–19.

Pankhurst, A. and Dom, C. (2019) *Rural Ethiopia in Transition: An Overview*. Available from: http://ethiopiawide.net/wp-content/uploads/WIDEBridge_Rural-Ethiopia-in-Transition-Overview.pdf

Pazarbasioglu, C., Mora, A., Uttamchandani, M., Natarajan, H., Feyen, E. and Saal, M. (2020) *Digital Financial Services*, Washington DC: World Bank Group.

Rahmato, D. (2004) 'Searching for tenure security? The land system and new policy initiatives in Ethiopia', Discussion Paper 12, Forum for Social Studies.

Rahmato, D., Zerihun, M., Webster, N. and Kefyalew, K. (2021) 'A scoping study of two localities in Ethiopia', *DIIS Working Paper*, Danish Institute for International Studies.

Rapsomankis, G. (2015) *The Economic Lives of Smallholder Farmers: An Analysis Based on Household Data from Nine Countries*, Rome: Food and Agricultural Organization, United Nations.

Ruhrort, L. and Allert, V. (2021) 'Conceptualizing the role of individual agency in mobility transitions: avenues for the integration of sociological and psychological perspectives', *Frontiers in Psychology*, 12: 623652. DOI: 10.3389/fpsyg.2021.623652

UNCDF (United Nations Capital Development Fund) (2021) *Inclusive Digital Economies for the Sustainable Development Goals*, New York: UNCDF.

Webster, N. (2023) 'Shaping spaces: governance and climate-related mobility in Ethiopia', *Climate and Development*: 1–11. DOI: 10.1080/17565529.2023.2227148

World Bank (2016) *World Development Report: Digital Dividends*. Washington DC: World Bank.

12

Conclusion: Policy Reflections on Slow-Onset Climate Mobility and the Importance of the Governance Variable

Neil Webster and Ninna Nyberg Sørensen

The Governing Climate Mobility (GCM) research programme, on which the chapters in this volume are based, was conceived in 2018–2019 at a time when considerable attention was being paid to the effects of climate change and to the challenges of migration pressure seen to be facing Europe and elsewhere. Early 21st-century optimism in migration development policy debates was increasingly being disregarded with a return to late 20th-century alarmist mass-migration scenarios. Insights from the Foresight Report on Migration and Environmental Change (2011) celebrated just a few years earlier, had argued that climate change only indirectly influences migration patterns, that it affects other existing drivers and that migration should better be understood as being an adaptation strategy. It also pointed to the fact that millions of people are not in a position to move away from those places in which they are severely vulnerable to the effects of slow-onset climate change. Such arguments were just a few years later being replaced with public and policy narratives that built on simple causalities along the lines of climate change causes mobility and international migration in particular. The latter had become the dominant concern for many.

What about governance, we asked? How might governance contexts and interventions influence and shape climate mobility practices? While the possible relationship between climate change and mobility practices had been examined before, the influence of governance as a significant factor prompting climate-related mobility had often been overlooked. Thus, we wanted to introduce a critical variable into any policy discussion, a variable that we saw as being largely absent. Consequently, we embarked on four years of research into the possible links between mobility, environmental change and the ways in which governance practices are shaping these.

With a theoretical and conceptual point of departure in the Foresight Report (2011) and key developments following its publication, in particular the climate mobilities and the aspirations and capabilities frameworks, two initial delineations were made. First, even if slow-onset climate change does not directly cause migration, it does exacerbate the vulnerability of individuals, households and communities living under precarious conditions. The focus on slow-onset climate change ensures that sudden changes causing involuntary migration, often of a temporary nature, have been set to one side for research purposes. This has enabled the GCM research to focus more narrowly on the agency of individuals and households in decisions on mobility and immobility, the influence of structural factors in shaping the contexts for such decision making and, most importantly the role of institutions of governance in shaping the contexts for such decisions. The last of these indicates the research programme's interest in the role of local, provincial and national institutions of government and that of other forms of political and traditional authority. For the GCM research programme, accepting the importance of governance institutions in contributing to the shaping of mobility practices in a particular context is to challenge the simple causalities behind the argument that climate change causes migration.

A second delineation was that mobility practices should be studied in their many forms. Distinctions based on the crossing of borders, on the temporary or permanent nature of the migration, on the period and purpose of the activity, on the characteristics of the households and individuals involved, and on the practice of immobility are but a few of the lines of enquiry taken up in the different chapters. Mobility and mobility practices became the terms adopted to ensure that a broad range of (in)voluntary (im)moblities are investigated in the research programme and the studies presented in this volume.

The four central questions that the GCM research programme addressed are as follows:

1. What are the historical dynamics of livelihoods and mobility in the study areas, and how and to what extent have they been affected by climate change and governance interventions?
2. How are livelihoods, mobility and climate change currently governed at the national and local levels, and what are the implications for households' and individuals' climate mobility options?
3. How and to what extent do households use mobility practices to adapt to climate change?

4. How do different governance contexts and practices at the local, national and international levels shape and frame households' mobility practices at the local level, and what are the implications of each for facilitating adaptive climate mobility?

These questions have guided the different studies presented in the preceding chapters and, in turn, have given rise to a range of considerations relevant for policy makers and practitioners. The purpose of this final chapter is to weave these into a set of insights that can shape and aid the policies developed and pursued on migration in a context of slow-onset climate change.

Slow-onset climate change is not the cause of mobility and migration practices

As has been noted throughout the volume, the argument that slow-onset climate change causes mobility and migration might be quite simple, but is seen as very plausible; it is an argument underwriting much that is driving public discussion and policy debates. Even where notions of 'complex causality' and 'migration as adaptation' have gained traction in select international policy circles such as the International Organization for Migration (IOM) and the Global Compacts on Safe, Orderly, and Regular Migration and Refugees respectively, the simple causality continues in government policies and programme design. In the case of African countries, the argument resides within much that is put into practice by national and subnational government authorities, by supporting donor organizations and by (international) nongovernmental organizations (NGOs).

When climate change is being discussed, the chapters in this volume have noted a strong tendency among Ethiopian and Ghanaian officials to link contemporary mobility quite directly and openly to slow-onset climate change. The logic of the argument extends further, mobility being often seen as a recent phenomenon and increasing rapidly, somehow in line with recent climate change. It is a discourse that sees those shaping policy striving to develop policies aimed at keeping people in place, presenting a professed desire to reduce human rights abuses linked to migration, and carrying a general wish to 'protect' vulnerable individuals and families. This is discussed in Chapter 7 in relation to practice and policies in northern Ghana and in Chapter 10's exploration of women's migration to the Middle East. At the same time, failures in recognizing the adaptive strategies being pursued by households, whether in the cultivation of khat (Chapter 6) or in short-distance migration (Chapter 8), results in counterproductive activities and missed opportunities by local authorities.

To develop a better and more nuanced perspective, those responsible for policy should re-evaluate the existing data, ask additional and different questions, and not least look to the history of mobility practised in the localities and communities that they are accountable for. There is also a need for greater sensitivity to the fact that slow-onset climate change affects similarly situated people in different ways and results in a variety of mobilities and immobilities, as discussed directly in Chapters 2, 3 and 8. Chapter 9 raises the importance of gender in addition to social and cultural contexts in shaping mobility practices and perceptions.

In accepting variation and multiple outcomes, one moves towards the acceptance of multiple causations being present and the need to address these. We therefore also suggest that any policy aimed at enabling people to stay in place in their households must consider the presence of well-established mobility patterns often dating back many years, and the extent to which these have long supported certain households' livelihoods. Research at the household level reveals mobility practices to be more often the rule rather than the exception. Two clear examples relate to pastoralism and to circular migration to in-country localities offering seasonal employment. Government authorities, international donor agencies and local NGOs need to ask themselves what role they play in these practices today, and how they might manage and modify their practices to support households better in the future.

Interventions must be sensitive to the intersecting dimensions of age, gender and other sociocultural hierarchies that in different ways shape who migrates, under what conditions and for the benefit of whom. See, for example, Chapter 10 and the discussion of women migrating to ensure education for their children or to access resource for building a house for their family or the movement 'south' in Ghana discussed in Chapters 4 and 7. Add to this the need to recognize the time dimension that any intervention needs to assess and address; for example, a food for work programme might need to have a predictable presence for more than a few months if the intended effect is to achieve long-term sustainable change for the households targeted and to affect long-established mobility patterns in any meaningful way.

Demographics, social networks and discourses are key considerations

Demographics are of key importance to understanding development dynamics, but although this might seem self-evident, it is often ignored in policy planning and implementation. As all the chapters have shown, demographics vary from locality to locality (see Chapters 2 and 3). Factors such as age, gender, class, education received and rural/urban location shape possibilities for climate adaptation and mobility practices (see Chapter 9). So, not having good demographic data is a serious weakness in any analysis of households' needs and aspirations, and seeking to work with and build upon these.

Taking this point a step further, it is important to recognize that individuals are embedded in social networks. Policy makers should therefore pay attention to how social networks and kinship ties facilitate and direct mobility practices in a particular locality and for specific social groups. Social networks based on family, friends and links to specific localities play several roles, including the provision of important financial support for the actual journey (for example, paying the agents involved), and are a potential base on arrival at the destination. And what of the immobile, those who 'stay behind'? Are they voluntarily or involuntarily immobile? Chapters 2 and 3 point to the importance of poverty reduction for reducing pressure on land and vegetation. Chapter 5 points to the failings of government in considering farmers' perceptions of climate change. Chapters 9 and 10 address the role of gender in affecting the ways in which mobility is perceived and practised. Chapter 11 argues the centrality of access to finance that can enable households to pursue improved livelihoods based on a broader range of options including mobility. Each one of the contributions in this volume in its own way points to real weaknesses to be found in government policies and planning, and in the interventions by other national and international agencies.

Social networks carry important narratives, often through social media, and these are present in the arguments made and the decisions taken by individuals and the households to which they belong. This raises the importance of discourses on and around mobility and immobility more generally, something that several of the chapters have highlighted. The stereotypical view that female migration is linked to sex work is found in both Ghana and Ethiopia (Chapters 9, 10 and 11). Discourses on migrants abound; they can be found in specific forums or linked to particular organizations, be they international or national in character. They can also be found in or directed at specific communities or social groups with identities rooted in age, gender, ethnicity, religion and so on.

The media coverage also plays a significant role in forming and presenting many of the dominant narratives found, magnifying the arguments and claims to which these often give rise. Addressing the potential for good and for harm present in the media's role should be a consideration in any intervention, with questions as to whose narrative prevails, based on what evidence, to what end and for whose benefit.

'Silo' thinking and interventions remain a challenge

As has been raised throughout the chapters of this volume and discussed earlier in this chapter, migration and mobility practices have often been analysed and interventions managed in ways that have been organized separately to work on slow-onset climate change. Governments and NGOs tend to have a sectoral approach to much of their work. Governance, education, health, agriculture and forestry can be seen as 'silos' in their organizational and institutional natures, in their budgets, and not least in their practices.[1] While sector-specific approaches can produce effective and efficient interventions, they can be counterproductive when seeking a more integrated approach particularly at subnational levels of government.

How might this be changed? One obvious step would be to think laterally across different sectors of government and across different fields of research. When working on and in a particular sector of development policy and practice, it would be relevant to introduce the following question: 'What about mobility practices – should we also engage with these?' The findings from the two household surveys, qualitative interviews and focus groups discussions conducted in Ethiopia and Ghana, on which the analyses presented in various chapters build, illustrate that respondents do not separate their children's' education, their families' illnesses, their land cultivation or other issues into compartmentalized areas of activity. These closely intersect and interact in their lives. Decisions in one area are closely bound to events and decisions taken in another. For example, why does a woman decide to migrate to Saudi Arabia to secure education for her children (see Chapter 10)? This intersecting of interests and actions needs to be reflected in policies and programmes.

A greater focus on the different tiers of local government would be an important step towards a more integrated approach. Strengthening their respective roles and resources could enable a more flexible intersectoral engagement with households' mobility and immobility practices. Local government is at the interface of citizens and the state in areas of service provision and economic interventions, in the identification of households' needs and aspirations, and in environmental management, involving several sectors and their local administrative resources. At the same time, local government is most effective when nationally endorsed and resourced, is implementing national policies and is regulated in their programme implementation by central government. Chapter 5 addresses the importance of securing greater equity in access to resources and for securing the necessary condition for households' adaptation to slow-onset climate change. Chapter 10 points to the problem of relying on occasional external funding for 'anti-trafficking' initiatives. Chapter 11 points to the importance of access to finance and the role of local government in facilitating this and national government in managing the market for finance.

Technical approaches produce technical solutions, not transformative change

The organization of development aid agencies and their activities, and of governments and the vertical nature of their administrative ministries serves to reinforce a tendency for policies to be implemented according to certain standard operating procedures. This can promote efficient and effective implementation, and can lead to improvements in a household's condition in absolute terms – for example, better school and health services, improved access to markets and to resources, efficient organization of social protection transfers. These can lead to a reduced need for household members to take seasonal work elsewhere or to travel abroad to seek employment. However, what it does not necessarily bring about is a change in a household's own ability to influence and shape its livelihood situation. This requires a rather different type of change in its situation, namely changes in its relations to other households, to those in positions of power and influence, and to those with control over key assets and resources such as land and credit. Chapters 4, 5, 6, and 11 all raise in various ways that policy might become more demand-driven and, in so doing, more responsive to household needs, thereby changing the nature of control and agency present.

The former absolute changes reflect a transactional approach to development and, in the research undertaken in the GCM programme, to working with mobility. The latter reflects a transformational approach through which exclusion is reduced, the possibility of having a 'voice' in policy design and implementation is increased, and a broader basis for household and individual agency is established.

To achieve a more transformational approach requires that households are engaged with and involved in decision-making processes, that the diversity of their experiences forms the basis for policy development and implementation, and that government and NGOs are more nuanced in approaching and learning from the mobility practices of rural households.

Possible improvements could include planning in intersectoral forums, governments bringing NGOs, international NGOs and other donors into common forums, and donors accepting that governments should support and facilitate, but not predetermine policy implementation by means of their resource provision. The resources made available to local government are a key factor. If grants are provided and local revenue collection is present, then these are unallocated and open to local decision making on their allocation and use. This is fundamentally different from allocated funds channelled down through specific sector ministries and departments, the so-called 'silo approach' discussed earlier. Accountability in their use can be regulated by central government while decisions as to emphasis and direction in their use can be driven more locally.

Research must challenge tendencies towards 'glocalization'

An underlying subtext to several of the chapters is that global narratives and perspectives are shaping the way in which the local practices among households and communities are seen and addressed. If the parameters and dynamics of this local are being determined by actors rooted in global narratives on migration, on climate change, on what constitutes a household's livelihood and on what drives individuals' aspirations, then the policies and interventions designed to address the vulnerabilities identified are unlikely to succeed. This globalization of the local – 'glocalization' (see Schuurman, 1994) – all too often ends badly for the households and individuals directly affected. Chapter 10 presents a clear example with the discussion as to how an emphasis on human trafficking surrounding women's irregular migration has serious consequences for the design and implementation of 'anti-trafficking' initiatives at the local level of communities and households. But, in fact, most of the chapters touch on similar tendencies, global perspectives shaping the understanding of local spaces with often detrimental consequences for households practising mobility as adaptation.

The GCM research has explored the practices of households and individuals in interviews, focus group discussions and structured surveys. The evidence has led to a critical questioning as to what constitutes evidence being used in support of government and other development actors' interventions. The research programme's findings strengthen a call to look at local evidence when addressing global narratives. To this end, the value of comparative and longitudinal (panel set) data in monitoring mobilities practised at subnational levels cannot be overstated. Similarly, qualitative evidence gathered from individuals, specific groups and those acting as agents for those migrating or staying in place gives depth and expression to the practices found. In addition, it is important to have the evidence that enables one to critically question the narratives used to justify a particular policy.

Government must work with households, not against them

A core finding throughout the volume is the need to consider mobility as a possible asset in slow-onset climate change adaptation and not to simply or simplistically view it as a problem. Once this finding is accepted, then it is clear that national and subnational government have central roles to play in aiding households' agency and slow-onset climate change adaptation. They also have a central role in supporting other organizations' work with mobility practices; these are the formal and nonformal authorities that several of the studies have explored.

While this reflection might sound obvious, it is a qualitative shift away from what has been found in the cases of Ghana and Ethiopia. It would require a major review of existing programmes, assessing their conducive or detrimental effects on mobility practices, and on their intended and unintended consequences. This would need to cover mobility in its many forms, including not only international migration, but also rural-urban migration and seasonal migration, as well as less obvious mobilities that might be daily, linked to long-distance trade networks and similar practices.

By way of an example, in current donor aid, the role of social protection has a significant presence. Yet few of the agencies involved consider mobility as an adaptation strategy that such programmes might indirectly facilitate, or for that matter the problems that these social transfers might also introduce. Whether old-age pensions, child education grants, food-for-work or cash transfers, these funds enter into the decisions taken by a household, and thereby its agency. Yet, the question of mobility is rarely addressed in the design, the targeting, the monitoring or the evaluation of such programmes. For example, see Chapter 10's call for programmes to look to those most likely to migrate and to enable better and safer choices to be made, and also Chapter 11's discussion of the role of different forms of finance in shaping various mobility practices found among the households studied.

A concluding comment

The studies undertaken for this volume are based on detailed data collected during several joint fieldworks undertaken in Ghana and Ethiopia just prior to the COVID-19 pandemic reached both countries and shortly before the outbreak of a civil war in Ethiopia. The data, both qualitative and quantitative, that have been collected are quite unique, as is the breadth of mobility practices taken up. While in many ways quite context-specific, the studies do open up a field of research that many already assumed to be well researched and well documented in publications.

The GCM research programme has questioned assumptions, approaches and practices, and has supported its work with strong field-based evidence. In so doing, the research-based studies have a relevance that extends far outside the localities in which fieldworks were undertaken. There is a strong resonance in the findings that touches directly on the work and policies of national governments and international organizations. They indicate quite clearly that to continue to work with migration as the problem and with slow-onset climate change as its cause does not offer researchers or policy makers a pathway towards sustainable solutions to the challenges that households and governments face.

The challenges presented by slow-onset climate change have been overlooked or underestimated for many years; however, today, not considering the significance of governance in shaping the agency of households as they face these challenges in their daily lives can only be seen as a failing on the part of researchers and policy makers.

Note

1. Sectors in government and policy are standard today and are reflected in development programmes that focus on education, health, forestry, agriculture and so on.

References

Foresight (2011) *Final Project Report: Migration and Global Environmental Change*, London: The Government Office for Science.

Schuurman, F. (1994) 'Agency, structure and globalization in development studies', in F. Schuurman (ed.) *Current Issues in Development Studies: Global Aspects of Agency and Structure*, Saarbrücken: Verlag fur Entwicklungspolitik Breitenbach, pp 12–55.

Index

References to figures and photographs appear in *italic* type; those in **bold** type refer to tables. References to endnotes show both the page number and the note number (249n2).

A

absolute poverty 198
abuse 210, 212, 214, 215, 219, 224, 225, 255
adaptation politics
   and climate-related mobility 150–155, 163–165
   and migration contestations 158–161
   reflections on 163–165
   see also irrigation schemes and initiatives
adaptive capacity 15, 89–90, 117
   adoption and institutions' role 91–92
   cross-scalar approach to 91–92
   data analysis by binary logistic regression models 94
   definition of 90
   livelihood adaptation and factors influencing 90–91
   quantitative and qualitative approach, using survey 93–94
   study areas in Ghana 92–93
   see also adoption of adaptation strategies
adaptive climate migrants 7
adoption of adaptation strategies 91, 94
   age and gender, household-level factors for 107–108
   agro-ecological zone location and household characteristics 107–109
   agronomic practices 97, **98–99**, 99, 101
   fertilizer application **100–101**, 101–102
   irrigation 94–97, **95–96**
   migration strategies 104, **105–106**, 106–107
   nonfarm activities 102, **103–104**, 104
afforestation 54–55
agricultural economy 27, 154
agricultural extension officers 117–118
Agriculture Based Individual Loan (ABIL) 245–246
agroforestry 54
agronomic practices 91, 97, **98–99**, 99, 101

agro-pastoralism 175
Amhara Region Agriculture Bureau 133, 142
anti-human trafficking taskforces 219
Arab Gulf 213–214
artisanal mining 47
aspiration-(cap)ability framework 5
autonomous adaption *see* indigenous adaptation

B

Bangladesh 235
belg season 56–58, 141–142
bilateral labour agreements 214
bimodal rainfall pattern 93, 193
binary logistic regression analysis 94–95, 102
biodiversity loss 24, 48, 54, 55, 80, 171
bush fallowing 47
Buusaa Gonofaa Micro-Finance Share Company (BGMFSC) 244–245

C

capacity-building funding scheme 11
catastrophic famines 144
cattle rearing 156, 174–175
Central Great Rift Valley region 171
Centre for Migration Studies (CMS), University of Ghana 11, 249n2
cereal production 55, 56–58, 140, 142, 174
charcoal production 46, 47, 48, 55
child labour 215
Chinese miners 47
circular migration 5
civil society organizations (CSO) 155
civil war, in Ethiopia 261
climate change adaptation strategies 115–116
   community strategies 117
   coping and adaptation practices 106, 117
   definition of 175–177

irrigation schemes as *see* irrigation schemes and initiatives
khat-centred adaptation strategy in Tehuledere 138–143
long-term and short-term strategies 116
migration as 153–155, 178–180, 191–192
off-farm strategies 116
on-farm adaptation measures 115, 116–117, 178
short-distance migration *see* short-distance migration
Climate Data Store (CDS) 63
climate hazards
government assistance due to 95–97, 99, 101–102, 104, 106
rainfall anomaly analysis 63, 66–71, *68, 69, 123*
Climate Hazards Group InfraRed Precipitation with Station data (CHIRPS) 63
climate mobility regimes 8
climate refugees 4, 7, 10, 150, 152, 165
climate resettlements 8
climate risk-assessment profile 171
climate variability 11, 90, 116, 151, 191, 197
adaptation strategies *see* adoption of adaptation strategies
and LULC changes 55–56, 79–81
political framing in response to 152–153
climate-related displacement 4–7
climate-related gendered migration 189–192
climate change perceptions 194–197
data collection 193–194
drivers of migration 197–201
migrants 189
rural-rural migration 204–205
rural-urban migration 203–204
study areas in Ghana 192–193
climate-related mobility 3, 6–10, 150, 253
and adaptation political framings 152–153, 163–165
adaptation strategies *see* adoption of adaptation strategies
gender dimensions of *see* climate-related gendered migration
governance perspective on 7–11
inclusive finance *see* inclusive finance
policy efforts to limit 151–152
politics and power in shaping 10
climatic shocks, sudden-onset 2
coffee plantations 55, 134, 243, 249
collaborative research programme 11
community leaders 117, 145
community tree planting activities 55, 81
community-based organizations (CBOs) 239–240
conservation 24
agriculture 54

of plant species 81
soil and water 80, 115, 116, 124, 178
cooperative loan 245
COVID-19 pandemic 217–218
impact on remittances 240–241
returned migrants during 211, 222–223
crop diseases 122, 128, 238
crop diversification 124, 126
crop production 48, 115–116, 155, 202, 240
adaptation approaches to 116–117
based climate change adaptation strategies 125–128, **127**
climate change impacts on 121–124
smallholder farmers *see* smallholder farmers
cropland 62, 72, 74, 76, 82, 115
cross-scalar approach 9, 90, 91–92, 109, 163

D

Danida Fellowship Centre 11
Danish Institute for International Studies (DIIS) 11
Danish Ministry of Foreign Affairs 11
Debub Wollo zone 56
deforestation 42–43, 46, 48–49, 55, 116, 171, 174, 175
delaloch (brokers) 220
detention camps 211, 221, 223
determinism 2, 5
Digital Elevation Model (DEM) 29
disaster risk management strategies 2, 6
displacement 4–5, 6, 29, 132, 134, 215
double cropping 58
double-maxima rainfall regime 93, 102, 107, 154
drought-resistant crops 126, 138, 140, 193
droughts 2, 114, 123, 128, 132, 134, 170–171, 197, 198, 200–202, 206, 216, 220, **240**, 242, 246
and adaptation strategies 125, 155
classification 67
and crop production 116, 121–122
increased number of 119
and private sector work 127
and soil water-holding capacity 80

E

economic push-pull factors 2
Economic Recovery Programme (ERP) 155
ecosystems 23, 141
degradation 24, 80
services 24, 54, 55, 115
ecotourism 48
Ejaza houses 222
enset (false banana) 174
environmental degradation 1, 3, 6, 80, 212, 219
classical form of 25
LULC changes lead to 55
theories explaining forms of 25

environmental hazards 114–115
environmental migration
  environmental refugees *see* climate refugees
  factors to 3–4, 5, 189, 193, 205, 233–234
  migrants 189
  studies 4
Ethiopia 3, 55, 114–115
  agro-ecological zones 56–58, *57*
  bias in economic development 249
  climate change on smallholders crop production *see* smallholder farmers
  data collection on migrants 217–219
  green economy plan 115
  homegrown economic reform (HGER) 115
  international migration rate 213
  migration for domestic work in 212, 213
  Shashemene district *see* Shashemene district, Ethiopia
  South Wollo zone of the Amhara region 238
  Tehuledere district *see* Tehuledere district, Ethiopia
  vegetation change detection analysis *see* vegetation change detection analysis, Ethiopia
  West Arsi zone, Oromia region 238
  women's irregular migration *see* women's migration, Ethiopia
eucalyptus 74, 79, 80, 82, 243, 249

F

Faji Sole (the highland kebele), Ethiopia 173, 174
Fanteakwa district, Ghana
  land cover changes in 31, *32*, **33**, 34
  location and agriculture 26, *27*
farming 91–93, 94, 125, 127, 133, 135, 140, 190
  adaptation 133
  agronomic practices *see* agronomic practices
  depletes forest cover 42, 46, 48
  dry-season 156, 162, 163
  households *see* households
  and irrigation 158–159
  migration during *see* godanttu (seasonal migration)
  from pastoralism to 174–175
  and rainfall regimes 201–203
  rain-fed *see* rain-fed agriculture farming
  season-based farming activities 141–143, 145
  smallholder *see* smallholder farmers
  tool fabrication 81
  see also nonfarm activities
faulty incentives 25
feminist scholarship 214
feminization 201

fertilizer application **100–101**, 101–102
floods 2, 34, 37, 80, 114, 116, 119–120, 132, 156, 162, 170, 174–175, 197, 200–203, 206, **239**, **240**
floristic composition and cultural importance, relationships between 81
focus group discussions (FGDs) 31, 93, 97, 107, 117–118, 119, 126, 179, 193–194, 217, 246
Food and Agriculture Organization of the United Nations (FAO) 46
food insecurity 137, 171
food scarcity 133
food security 81, 115, 123, 126, 202
food-deficit area 133
forced immobility 5
forced prostitution 215
foreign direct investment (FDI) 240
Foresight
  cross-scalar approach 92
  Report on Migration and Environmental Change (2011) 253, 254
forest vegetation 31
Forest Zone (Eastern region), Ghana 26–28, 89, 189
  adoption strategies *see* adoption of adaptation strategies
  changes in temperature and rainfall 190, 194–197
  climate change as drivers of migration 197–201
  deforestation in 42, 46, 48–49
  patrilineal and matrilineal inheritance 190, 201
  study areas for adaptive capacity 92–93
  see also Fanteakwa district, Ghana; Yilo Krobo district, Ghana
formal banking services, availability and accessibility 237
formal state institutions 7–8
Forum for Social Studies (FSS), Ethiopia 11
fuelwood 47, 55, 80, 81
funds transfers 138, 211, 237, 243, 247–248, 259, 261

G

galamsey, illegal surface mines 154, 159, 205
galessa, khat variety 138
GCM Household Survey 134–135, 136, 143, 146
gender 6, 9, 91, 93, **95**, **98**, **100**, **103**, **105**, 109, 164, 172, 178, 256, 257
  gender-based violence 214, 216–217, 219, 225
  gendered farming practices 97
  gendered mobility practices *see* climate-related gendered migration
  influences to adopt irrigation 108

trespassing boundaries *see* women's migration, Ethiopia
gender dimensions of climate-related mobility 197
geographic information system (GIS) 11, 12, 24, 28, 31, 56
Ghana 3
　adaptive capacity assessment *see* adaptive capacity
　agro-ecological zones 26–28
　Eastern region *see* Forest Zone (Eastern region), Ghana
　Ghana Agricultural Development Programme (GADP) 156
　north-south labour migration 156
　political ecology perspective on vegetation loss in 25
　politics of adaptation and climate (im)mobility 150–153
　southern part of 154–155
　Upper West region *see* Northern Savannah zone (Upper West region), Ghana
　vegetation change detection analysis *see* vegetation change detection analysis, Ghana
Ghana Statistical Service (GSS), 2018 190
Global Compacts on Safe, Orderly, and Regular Migration and Refugees 255
Global Positioning System (GPS) 136
glocalization 259–260
godanttu (seasonal migration) 180
Google Earth Engine (GEE) 58
governance, scalar dynamics of 7, 8–9
governance perspective on climate mobility
　differentiation across social groups 9–10
　plurality in governance actors 7–8
　and political authorities' role 7
　scalar dynamics of 8–9
　structure-agency interplay 10
government assistance, to households 95–97, **99**, 101–102, 103–104, **106, 240**
government-led inclusive finance 236
green cover
　classification type, Ethiopia 61–62, **62**
　classification type, NDVI 30, **31**
　rate and change over time 24, 28–29, 58, 82
green economy plan 115
greenhouse gases 114
greenness index, NDVI-derived 28
group loan 245–246

H

habitat fragmentation 24
handicrafts **125**, 243, **244**
hierarchies of exploitation 237, 248
home-gardening practices 81
homegrown economic reform (HGER) 115

household 4, 5, 6, 194, 198, 201, 203, 254–261
　adaptive capacity *see* adaptive capacity
　assets 132, 134–135
　characteristics of 107–109, 118–119
　decision making in mobility *see* households, decision making in mobility
　khat plots allocation *see* khat cultivation
　migration *see* adaptation politics; short-distance migration
　returned migrants 210–211, 217–219
　smallholder farmers *see* smallholder farmers
households, mobility decision making
　adaptation strategy for 243, **244**
　challenges faced by 238–239, **239**
　factors influencing 233–234, 254
　finance, access to *see* inclusive finance
　government assistance and accessing to credit 239–241, **240, 241**, *241*, 260–261
　land and land rights 242–244
　mobility practices *see* mobility practices
human rights abuses 255
human smugglers 211–212, 219, 224
human trafficking 14, 210, 212–213, 214–215, 216, 218–220, 221, 224, 260
　see also irregular migration
hydrological droughts 80

I

illegal chainsaw operators 46–47
illegal mining 47, 154–155, 158, 205
immobility hierarchies 4
inclusive finance 15, 236–237, 248–249
　data collection from GCM field sites 237–238
　digital 247–248
　formal banking services, availability and accessibility 237
　land and land rights 242–243
　loans by micro-finance institutions 245–247
　micro-finance, access to 244–247
　and mobility dynamics 234–235
　social benefits 242
　transactional change *see* funds transfer
India 55
indigenous adaptation 132–133, 146
inequality 198, 240, 247–248
informal governance actors 8
inheritance rights 135–137
institutional credit 235, 236
insurance, crop and cattle 124, 125, **244**
Intergovernmental Panel on Climate Change (IPCC) 90, 91, 189, 197
International Finance Cooperation (IFC) 241
international migrants 178
International Monetary Fund (IMF) 155
International Organization for Migration (IOM) 222–223, 255

INDEX

investments 154, 211, 223, 237, 247, 249
  FDI *see* foreign direct investment (FDI)
  in public programme 137
  of remittances 243
  security 242
  for urbanization 133–137
Iraq 213–214
irregular migration 14, 47, 210, 211–217, 219–220, 221, 260
irrigation schemes and initiatives 109, 116, 150–151, 163–166
  adaptation as multiscalar sociopolitical process in 157–158
  adoption of 94–97, **95–96**, 108–109, 137, 140, 189, 202
  construction of 137, 140
  development 155–156
  differences in generational perspective on 158–160
  negative impacts of 153
  One Village-One Dam (1V1D) initiative 156–157
  outcomes of 160–163
  in situ irrigation 155–158

J

Jirapa district, Ghana 154
  adaptation strategies *see* adoption of adaptation strategies
  land cover changes in 38, *39*, *40*, **41**, 42
  location and agriculture 26–28, *27*
  study areas for adaptive capacity 92
Jordan 213–214

K

kafala system 213–214, 219, 220
Karakara, A. 247
kebele
  in Shashemene, Ethiopia 172–175, 179, 180, 217, 218, 238, **240**, 246
  in Tehuledere, Ethiopia **134**, 134–135, 136, 143–144, **144**, 145
khat cultivation 138–140, 249, 255
  adaptation strategy in Tehuledere 140
  climate change impacts on smallholders 141–143
  debate on 139–140
  khat growers 132, 134, 135, 137
  khat-chewing seasons 145
  market-driven 145–146
  plot allocation to households 143–145

L

Labour and Skills Office, Shashemene 210
land cover changes 24
  Fanteakwa district, Ghana 31, *32*, *33*, 34
  Jirapa district, Ghana 38, *39*, *40*, **41**, 42

LULC *see* land use and land cover (LULC) change
  positive and negative consequences 79–81
  Shashemene district, Ethiopia 74, 76, *76*, **77**, *78*, 78–79, **79**, 210, 217, 218
  shrubland/TOF expansion and woody vegetation cover increase 80
  Tehuledere district, Ethiopia 71–72, *72*, **73**, 74, 218, 220
  Wa West district, Ghana 42, *43*, *44*, **45**
  Yilo Krobo district, Ghana 34, *35*, *36*, *37*, 38
land degradation 55, 79–80, 116, 171
Land Reform (1975) 175
land rehabilitation programmes 80, 81
land use and land cover (LULC) change 24–25, 54, 193
  and climate variability 55–56
  deforestation 55
  detection using GIS and remote sensing 56
  leads to environmental degradation 55
  vegetation change detection analysis *see* vegetation change detection analysis, Ethiopia
Lebanon 213–214
livelihood adaptation 90–91, 92, 202
livestock 124, 193
  diseases 119, 132
  production 55, 92, 115, 126, 134
local authorities 117, 220, 255
logging 25, 46–47, 49
long-gestation crop 123, 126

M

maladaptation 2, 9, 150, 162–163, 166
Mann-Kendall test 63–64, 66
marginal lands 174, 175
marginalized populations 9, 163, 245
matrilineal inheritance 190, 192, 201
meher season 56–58, 141–142
mengdemerwecho 220
micro-finance 235–236, 244–247
migration for domestic work in Middle East, Ethiopia 213
  human trafficking 214–215
  kafala system *see* kafala system
  violence, alternative to various forms of 215–218
migration governance 6, 214
migration labour laws 225n4
migration strategies, adoption of 104, **105–106**, 106–107
migration theorization 3–4
migration-crime-security-nexus 215
mining 25, 46, 47, 49, 158, 205
Ministry of Food and Agriculture (MoFA) 157
mixed cropping 97

mobilities and immobilities,
  relationship between 9
mobility practices 3, 5, 7, 9, 91, 215, 243,
  253–254, 258–261
  dynamics of 234–235
  funds transferring ability *see* funds transfer
  gender dimensions *see* climate-related
    gendered migration
  politics *see* adaptation politics
  and slow-onset climate change hazards 235
  and social networks 256–257
model farmers 245–247, 249–250n10
modernization opportunity 198, 201

N

natural resources
  exploitation of 170
  loss of 137, 173
natural vegetation 12, 23–24, 80
negative rainfall anomaly 66–67
neoliberal perspective, on vegetation cover
  loss causes 25
non-agricultural income 55, 124
nonfarm activities 102, **103–104**, 104,
  243, 249n10
nongovernmental organizations (NGOs) 117,
  125, 162, 217, 218, 236, 239–240,
  255, 256
  helping women in adopting irrigation
    96–97, 108
  and micro-finance 235, 245
  providing loans to set up shops 211
  sectoral approach 257–258
  transformational approach 259
non-timber forest products (NTFP) **120, 239**
Normalized Difference Vegetation Index
  (NDVI) 28–29, 58, 61
Northern Development Authority (NDA)
  see Savanna Accelerated Development
  Authority (SADA)
Northern Savannah zone (Upper West
  region), Ghana 26–28, 89, 153, 189
  adoption strategies *see* adoption of
    adaptation strategies
  climate change and migration in 153–155
  climate change as drivers of
    migration 197–201
  irrigation schemes development 155–158,
    160–163
  migration contestation 158–160
  Northern Savannah Ecological Zone
    (NSEZ) 156
  patrilineal inheritance system 190, 201
  perceptions of temperature and rainfall
    in 194, **195, 196,** 197
  rapid vegetation cover loss in 46–47
  Savanna Accelerated Development
    Authority (SADA) 156

study areas for adaptive capacity 92
  see also Jirapa district, Ghana; One
    Village-One Dam (1V1D) initiative;
    Wa West district, Ghana
Northern War 135

O

off-farm adaptation strategy 94, 102, 116,
  178, 180, 183
off-forest tree coverage 55, 81
Office to Monitor and Combat Trafficking in
  Persons 215
One Village-One Dam (1V1D) initiative 13,
  156–157
on-farm adaptation measures **103–104,** 115,
  116–117, 178
organ harvesting 215
Oromia Credit and Savings Share Company
  (OCSSC) 244–246
overseas development assistance (ODA) 240

P

pastoralism 174–175, 180, 256
  see also cattle rearing
patriarchy 14, 91, 201
patrilineal inheritance 190, 192, 201
pest infestation 132
pharmacopoeia 48
policy development and implementation
  258–259
political ecology 9, 24, 25
political economy 25
polygamy 216
positive rainfall anomaly 66–67
power dynamics 4, 6, 10, 153
private moneylenders 235
Productive Safety Net Programme (PSNP)
  242, 243
property rights 25, 136, 242, 243

Q

Qatar 213–214, 225n4
quarantine 211, 222, 223

R

racial discrimination 214
rain-callers 198
rainfall anomaly analysis 63, 66–71, *68, 69, 123*
rain-fed agriculture farming 114, 122,
  126, 128, 151, 153–154, 161, 170, 192,
  193, 198
reforestation 54–55
rehabilitation 54–55, 80, 81, 211, 218
remittances
  investment of 243
  migrant 2, 150, 159
  transfer of 237, 240–241, 247
  women's 214, 222

resettlement 5, 8
return movement 5
returned migrants 210–211, 217–219
rich households 246
river blindness 48
row planting 137
rural development initiatives 137
rural-rural migration 160, 198, 204–205
rural-urban migration 104, 156, 160, 198, 203–204, 245, 261
  long-distance migrants 178–179
  short-distance migrants 171, 178–183

S

satellite sensor characteristics 29, **30**, **61**
Saudi Arabia 179, 214, 220, 221, 223, 258
Savanna Accelerated Development Authority (SADA) 156–157, 166n2
seasonal mobility/migration 5, 124, **125**, 154, 162, 175, 180, **198**, **199**, 200, 206, **244**, 261
seasonal Sens's slope 63–64, 66
Sen's slope 63–64, 66
shades, selling spot 210–211
Shashemene district, Ethiopia 55, 117, 171, 217
  change detection analysis image datasets 58, **61**
  climate change adaptation 124–125, 175–178
  climate change and perceptions in 173–174
  climate change impacts on crop production 119, **120**, 121, *122*, *123*
  crop production-based climate-change adaptation strategies 125–128, **127**
  kebeles of 173
  land cover changes in 74, 76, *76*, **77**, *78*, 78–79, **79**
  livelihood dynamics of 174–175
  location and climate of 56, *57*, 58, *60*, 172–173
  migration in 178–180
  pastoralism to agro-pastoralism, shift from 175
  rainfall and temperature trend 64, *65*, 66, *66*
  rainfall anomaly and CV of 67, *68*, **70**, 71, **71**
  rural-urban migration *see* rural-urban migration
  smallholder crop production *see* smallholder farmers
  trend analysis of temperature and rainfall 121, *122*, *123*
short rotation crops 123, 126
short-distance migration 171–172
  climate change adaptation and rural-urban migration 175–180

climate change and perceptions 173–174
  from pastoralism to farming 174–175
  socioeconomic implications of 180–183
shrub/trees outside forests (TOF) *see* trees outside forests (TOF)
silo approach, challenge of 257–258
single maxima rainfall regime 106, 202
single rainfall regime 27, 154, 155
Sinqe Bank 245
Siriyiri, Wa West District 204
slow-onset climate change 2, 4, 15, 233, 235, 238, 253–261
  adaptation 260–261
  direct effects of 238
  involuntary migration 254
  and migration practices 255–256
smallholder farming *see* smallholder farmers
smallholder farmers
  adaptation strategies of 140
  adaptation strategies to climate change 116–117, 124–125, **125**
  characteristics of 118–119
  crop production-based climate-change adaptation strategies 125–128, **127**
  farming 133, 233
  partnership with private sector 127–128
  perceptions of climate change 119, **120**, 121, *122*, *123*
social benefits 242, 243
social inequality 6
Social Protection Policy of 2014 242
soil and water conservation 80, 81, 115, 116, 124, 178
soil degradation 55, 79–80, 116, 161, 171, 193
soil erosion 48, 55, 80, 123, 174, 176
soil fertility 47, 101, 119, 175, 201, 202
soil impoverishment 101, 198
soil infertility 202
soil water-holding capacity 80
SORRETI 127–128
South Wollo zone, Amhara region 132–135, 137–139, 141, 211, 217, 238–243, **244**, 247
structure and agency, relationship between 5, 10
Sub-Saharan Africa (SSA) 114, 119, 170, 213
sustainable forest management practices 54

T

Taungya system 26
teff 58, 118–119, 123, 124, 126, 127, **127**, 133, 238
Tehuledere district, Ethiopia 54, 55, 132, 217–218, 220
  adaptation strategies of smallholder farmers 140
  agro-ecology zones 141–142, *143*
  belg and meher, agricultural seasons 141, *142*

change detection analysis image
  datasets 58, **61**
climate change 141–143
consequences of land cover changes in 80
debate on khat cultivation 139–140
kebeles of **134**, 134–135, 143–144, **144**
khat cultivation in 138–139
khat plots allocation to
  households 143–145
land cover changes in 71–72, *72*, **73**, 74,
  75, *75*
land tenure, registration and
  certification 135–137
location and climate of 56, *57*, 58, *59*
market-driven khat cultivation 145–146
rainfall and temperature trend 64, *65*,
  66, *66*
rainfall anomaly and CV of 67, *69*, **70**,
  71, **71**
rural development initiatives and public
  programme implementation 137–138
rural livelihoods in 133–138
urban economy and consumer goods
  of 134–135
urbanization 133–135
timber 23, 24, 26, 46, 47, 48
timely planting 126
topographic displacement 29
traffickers 211–212, 215, 220, 224
transactional changes 247, 248, 259
transformative change 248–249, 258–259
tree plantation 81
trees outside forests (TOF) 55–56, 62–63,
  75, *75*, *78*, **79**, 79–81
trend analysis, rainfall and temperature 63–66,
  65, *66*, 82, 121, 194, 197
trespassing, boundaries 212–213, 217, 221,
  223–225
  see also migration for domestic work
  in Middle East, Ethiopia; women's
  migration, Ethiopia

U

United Arab Emirates 214
urbanization 3, 133–136, 154–155
urban-rural migration 104
US Geological Survey (USGS) 29

V

vegetation change detection analysis,
  Ethiopia 54–56
  approach and data for 58, 61, **61**
  climate data analysis 63–64
  data pre-processing 61
  land cover change in Shashemene and
    Tehuledere district 71–79
  location and climate in study areas 56–58,
    59, *60*

NDVI computation and image classification
  61–63, **62**, **63**
rainfall and temperature trend
  analysis 64–66
rainfall anomaly for climate hazards 66–71
vegetation change detection analysis, Ghana
  causes and effects of vegetation cover
    change/loss 42, 46–49
  data and pre-processing of **29**, 29–30, **30**
  Eastern region land cover changes 31–38
  interviews and FGDs with local
    communities 31
  location and study regions 25–26, *27*, *28*
  NDVI computation and image
    classification 30, **31**
  Normalized Difference Vegetation Index
    (NDVI) approach with GIS 28–29
  Upper West region land cover changes
    38–42, *43*, *44*, **45**
vegetation cover change/loss
  causes and effects of 42–49
  conversion rates 23
  by deforestation 42, 46, 48–49
  detection analysis *see* vegetation change
    detection analysis, Ethiopia; vegetation
    change detection analysis, Ghana
  due to logging 46–47
  ecological and socioeconomic impacts of 48
  high rates of 24
  land use and land cover (LULC)
    change 24–25
  by mining 47
  poverty and population increase causes 47
  theoretical perspectives on 25
  by wood cutting and road, settlements
    construction 47

W

Wa West district, Ghana 154
  adaptation strategies *see* adoption of
    adaptation strategies
  climate change perceptions 194–197
  gendered and climate-related forms of
    migration 197–203
  land cover changes in 42, *43*, *44*, **45**
  location and agriculture 26–28, *27*
  rural-rural migration 204–205
  topography of 192–193
water scarcity 48
West Arsi zone of the Oromia region 56,
  117, 172, 217, 238
WIDE programme 136
Women and Children's Affairs regional
  office 211
women's migration, Ethiopia 14, 215–217,
  219–220, 225
  as capability-enhancing choice 222
  employers or relatives, deceived by 221

human trafficking and labour
  exploitation 220
 return during COVID-19 222–223
 trespassing legal boundaries 221
woody vegetation 56, 72, 82
 cover changes 62, 78, 80
 in Shashemene district 76, 78,
  78–79, **79**
 in Tehuledere district, Ghana 72, **73**, 74,
  75, 75
Worobong forest reserve 26

Y

Yilo Krobo district, Ghana
 adaptation strategies *see* adoption of
  adaptation strategies
 climate change perceptions 194–197
 gendered and climate-related forms of
  migration 197–203
 land cover changes in 34, *35*, *36*, **37**, 38
 location and agriculture 26, *27*, 92–93
 topography of 192–193
 youth migration 154–155, 158–160

www.ingramcontent.com/pod-product-compliance
Lightning Source LLC
Chambersburg PA
CBHW070803040426
42333CB00061B/1875